Diversity in Women's Entrepreneurship

The focus of this book is on women-led initiatives and innovative solutions for women in entrepreneurship. It sheds light on the unique challenges faced by diverse groups of women from various fields and how their collective contributions drive sustainable development globally, specifically in African countries.

Diversity in Women's Entrepreneurship: Global Sustainability in African Countries examines the professionalism and expertise of women in innovation and entrepreneurship, recognizing and analyzing the varied experiences of women from diverse backgrounds. It presents cases and accounts of women in African countries, emphasizing how empowering women across various intersections can lead to more impactful sustainable development on a global scale.

This professional book takes a multidisciplinary approach, integrating business, engineering, and sustainability as it provides a comprehensive understanding of women's entrepreneurship in various domains.

Women in Industry 4.0: Shaping World Foundations in Engineering, Business, and Sustainability

Series Editors: Tilottama Singh, Richa Goel, and Kayla L. Tennin

According to a World Economic Forum report, only 22% of AI professionals globally are female. This series will bring stories straight from the heart of these innovative women who have established themselves, to inspire millions and create a wave of impact. This will be a series of books and authors that will focus on sustainability through technology and entrepreneurship in Industry 4.0. The series will act as a global platform for bringing together organizations and women sharing their life cases, research, and grounded theory which will be an interesting read for engineers, businesses, government, researchers, and students working on startups and contributing to gender diversity.

Each book in the series is dedicated to illuminating women's research and achievements in key, targeted areas of the contemporary tech world and raising awareness of the pivotal work women are undertaking in areas of importance to our global community toward sustainable development. It will be a step in bringing forth the reality of what overshadows women in workplaces and the strategic attempts of organizations to pave the path for women's representation in companies and fields of engineering.

Books that focus on women in engineering and technology exploring and describing how women have greatly contributed to the world at large through these fields and their many subsections is a focal point of this series.

If you are interested in writing or editing a book for the series or would like more information, please contact Cindy Carelli, cindy.carelli@taylorandfrancis.com.

African Women Trailblazers in STEM
Pioneering Technological and Economic Advancement
Edited by Uzo Mkparu, Richa Goel, Tilottama Singh and Jade Mkparu

Women, Tech, and Enterprise
A Path to Sustainable Equality in South Asia
Edited by Kavita Indapurkar, Neeru Sidana, Richa Goel, and Tilottama Singh

Tech Startups Driving Innovation
Empowering Women in Bangladesh
Edited by Quazi Tafsirul Islam, Richa Goel and Tilottama Singh

Diversity in Women's Entrepreneurship
Global Sustainability in African Countries
Edited by Bola Adijatu Yaya, Elijah U. Felix, Tilottama Singh and Richa Goel

Diversity in Women's Entrepreneurship

Global Sustainability in African Countries

Edited by Bola Adijatu Yaya, Elijah U. Felix, Tilottama Singh, and Richa Goel

CRC Press
Taylor & Francis Group
Boca Raton London New York

CRC Press is an imprint of the
Taylor & Francis Group, an **informa** business

First edition published 2026
by CRC Press
2385 NW Executive Center Drive, Suite 320, Boca Raton FL 33431

and by CRC Press
4 Park Square, Milton Park, Abingdon, Oxon, OX14 4RN

CRC Press is an imprint of Taylor & Francis Group, LLC

ISBN: 978-1-032-83337-8 (hbk)
ISBN: 978-1-032-83359-0 (pbk)
ISBN: 978-1-003-50895-3 (ebk)

DOI: 10.1201/9781003508953

Typeset in Times
by Apex CoVantage, LLC

Contents

Foreword

Sustainability and innovation guide the modern world as it transitions toward business and development models of the future. The continent of Africa contains abundant resources and cultural heritage alongside exceptional human resources while nurturing motivated women who reshape both industrial standards and economic perspectives. This timely research looks at the fundamental part performed by women in African business innovator and sustainability positions.

As an advocate for gender equality combined with economic development, I have been allowed to see firsthand the strong determination and creativity of African women operating in business and technology fields. Women across Africa establish market positions within areas that were previously reserved off-limits through their innovation in renewable energy projects, along with engineering advancements, fintech developments, and sustainable community initiatives. The book shares riveting stories about the dynamic achievements and challenges faced by women in African business and technology.

These case studies with analytic backgrounds serve organizations led by policymakers as well as researchers and business executives as important reference sources. Youthful women and upcoming businesspeople will draw motivation from these studies to transcend constraints and construct Africa's economic and technological progression.

This reading experience will enable you to evaluate current achievements while understanding the remaining objectives. This book should inspire action to support women through mentoring and upliftment initiatives that benefit sustainable business and innovation. In unison, it is possible to create an environment in which abilities are boundless by gender, and discriminating against female talent should not be the norm.

Dr Sujata Shahi
Vice Chancellor of IILM University
Gurugram, Haryana, India

Preface

The global business landscape defines itself through sustainability and innovation through the transformative leadership of African women in this current era. Despite systematic challenges across all industries, from sustainable businesses to engineering, technology, and leadership sectors, women are pushing those lines. What African women are doing as entrepreneurs, as engineers and innovators, and as transformational leaders, purposefully promoting sustainable business approaches, is featured in the book.

This book introduces a number of topics about sustainable business for women alongside technological advances, leadership development, and relevant practical case examples, which offer a background about the modern female influence on African economic and social progress. This collection is a tribute to women and a source of inspiration for future generations of women in male-dominated professions.

The goal of this book is to make visible success stories so that policymakers, educators, and business leaders can use a sequential plan to create environments that support female innovation and sustainable development. This publication illustrates the possibility of leading through resilience and creativity that helps build a sustainable path toward inclusivity on the part of African women.

Editors

Bola Adijatu Yaya is a training and research lead with over five years of multifaceted experience. She is a cross-functional professional and a trailblazer at the intersection of gender equality, social inclusion, technology, and research. As a proactive advocate for women's leadership and empowerment, Yaya is the research lead at the Centre for Women Inclusion and Strategic Development and has lent her expertise to international nongovernmental organizations (INGOs), spearheading projects to break down barriers and elevate women in leadership roles. Her initiatives encompass promoting diversity, fostering inclusive workplaces, and providing crucial training resources for women's career success. She is a published research scholar venturing into entrepreneurship as the co-founder of Eloquent Data, a groundbreaking data analytics learning startup, where she also serves as the head of marketing and communications. Through this endeavor, she is also on a mission to democratize access to data analytics education, particularly prioritizing affordability for young professionals, with a special emphasis on empowering women in the field. She has been nominated for multiple prestigious awards locally and internationally and is a member of well-known professional communities around the globe, such as Forbes and AIESEC.

Elijah U. Felix believes firmly in the inclusion of youth and women in leadership and decision-making; thus, he works with different organizations, both locally and internationally, focused on youth and women's development. He loves connecting with other young individuals around the world who are involved in civic activities in their local communities and has participated in several youth leadership conferences, including Model United Nations General Assemblies.

He is a member of some global organizations, such as the Junior Chamber International (JCI) and AIESEC. He has a bachelor of engineering degree from the University of Benin, Nigeria, and an executive (mini) MBA in leadership and management from the Accra Business School, Ghana. He is a JCI Nigeria-certified trainer and facilitator.

Elijah U. Felix has conducted several leadership trainings and facilitated learning sessions both within and outside Nigeria. He is an associate member of the Institute of Strategic Management of Nigeria and The BoardRoom Institute of Ghana.

Currently, he is Executive Director for the Centre for Women Inclusion and Strategic Development, a feminist organization that empowers women in leadership. Also, he is a financial trading expert and a fellow of the African Young Leaders Fellowship Program in Ghana. He is undergoing a fellowship program in the 12th Set of the African Centre for Leadership Strategy and Development in Abuja while building a brand in the knowledge industry as a corporate leadership facilitator.

Dr. Tilottama Singh is a prolific author of books and series on sustainability. Her recent books and webinars, along with sessions on poverty, sustainable technology, women entrepreneurs, and Metaverse, with international publishers gained a great response through media coverage, international conferences, and global entrepreneurship conclaves at the global level organized under her leadership as Head of Department, School of Management, Uttaranchal Institute of Management, Uttaranchal University, Dehradun. A certified HR analyst from IIM and a proficient academic, researcher, and trainer with more than 15 years of experience, Dr. Singh is the recipient of the 2023 State Award by the government for her exemplary contribution in the field of sustainability and community development activities at the institution level. Her areas of expertise are human behavior and emotions and psychological testing. She has an enriched research portfolio with Scopus and international refereed journals, books, and conferences and holds 11 patents to her credit. A seasoned International Executive Development Program trainer, she has organized two consecutive EDPs in the past two years for government officials from the Republic of Tanzania in the field of price capping for poverty alleviation.

Dr. Richa Goel is an accomplished academician with 23-plus years of experience in the field of economics and management. Having expertise in diversity management for six years during her PhD cycle gave her a lot of insight, which she has reflected in her published works and trainings. Her projects on E-Shiksha, women empowerment, and inclusive banking using the Mahdia model gained recognition from the Ministry of Finance, New Zealand. She has worked very closely in areas related to sustainable economies, economic modeling, development economics, demography, public finance, SDGs, economic development, and international business in emerging markets. Dr. Goel has published numerous books and global case studies. Working in the area of women empowerment and leadership for many years, Dr. Goel has connected with a large global audience, including researchers, academicians, stakeholders, and industry guests, by creating more space for speaking platforms like trainings, seminars, or FDPs, along with conducting many international and global conferences like INCONSYM'2024 (Symbiosis University), IEEE/INBUSH/IB SUMMIT/Global Case Studies Summit (Amity University) with respect to SDGs. She has been handling more than four book series as Series Editor in areas of women empowerment, leadership, entrepreneurship, and e-commerce transforming societies in accordance with UNDP SDGs.

Contributors

Saumya Agarwal
Research Scholar
Harcourt Butler Technical University
Kanpur, Uttar Pradesh, India

Yukti Ahuja
Professor
JIMS
Delhi, India

Aanchal Amitabh
Adjunct Professor
Master School of Management
Meerut, Uttar Pradesh, India

Mildred Amugune
Researcher
African Leadership University
Nairobi, Kenya

Sukanta Kumar Baral
Dean and Professor
Faculty of Commerce & Management
Indira Gandhi National Tribal University
Amarkantak, Madhya Pradesh, India

Sonam Bhadauriya
Assistant Professor
Swami Rama Himalayan University
Dehradun, Uttarakhand, India

Sheela Bijlwan
Assistant Professor
Swami Rama Himalayan University
Dehradun, Uttarakhand, India

Aissam Bouadjel
Position Health, Safety, Environment & Quality Head (Director)
Fujairah Oil Industry Zone (FOIZ)
Fujairah, United Arab Emirates

Sheetal Chadda
Professor
JIMS
Delhi, India

Madhusmita Dash
Associate Professor
Siksha O Anusandhan (Deemed to be University)
Bhubaneswar, Odisha, India

Franklin Easter Dumbari
President and Chancellor
Franklin International Business School
Rivers State, Nigeria

Vikas Garg
Symbiosis Centre for Management Studies
Symbiosis International University (Deemed to be University)
Noida and Pune, India

Abdulsattar Abdullah Hamad
Professor
University of Samarra
Samarra, Iraq

Hrishikesh Kakde
Assistant Professor
MGM University
Chhatrapati Sambhajinagar Maharashtra, India

Ali Haj Khalifa
Associate Professor
College of Business Administration
University of Khorfakkan
Sharjah, United Arab Emirates

Nik Zulkarnaen Khidzir
Deputy Director, Associate Professor
UKM International
Faculty of Creative Technology and Heritage
University Malaysia
Kelantan, Malaysia

Chitra Krishnan
Symbiosis Centre for Management Studies
Symbiosis International University (Deemed to be University)
Noida and Pune, India

Amit Kumar
GL Bajaj Institute of Management and Research
Greater Noida, Uttar Pradesh, India

Arpana Kumari
Assistant Professor
Symbiosis Centre for Management Studies
Symbiosis International University (Deemed to be University)
Noida and Pune, India

Sasmita Nayak
Principal
Koustuv Business School
Bhubaneswar, Odisha, India

Rotimi Adeniran Owoade
Honeybrook Investments Inc
Toronto, Canada

Pallavi Padhy
Research Scholar
Siksha 'O' Anusandhan (Deemed to be University)
Bhubaneswar, Odisha, India

Mousumi Panda
Professor and Course Head
Centre for Agri Management
Utkal University
Bhubaneswar, Odisha, India

Rashika Panday
Chief Ecosystem Officer
Quantum EdgeX
Johannesburg Metropolitan Area
South Africa

Gary W. Paul
Chief ESG Officer and HR Director
Enfin Energy Finance
Bloemfontein Metropolitan Area
South Africa

N.D.N.B. Rathnayake
PhD, University of Colombo
Colombo, Sri Lanka

Kharabela Rout
Assistant Professor
Entrepreneurship Development
KISS Deemed to be University
Bhubaneswar, Odisha, India

Vijaya Rudraraju
Professor
GIET University
Gunupur, Odisha, India

Supriya Lamba Sahdev
Alliance School of Business
Alliance University
Bengaluru, Karnataka, India

Priti Ranjan Sahoo
Associate Professor
KIIT Deemed to be University
Bhubaneswar, Odisha, India

Pravinkumar Govind Shastri
Associate Professor
MGM University
Chhatrapati Sambhajinagar, Maharashtra, India

Anuja Shukla
Associate Professor
Jaipuria Institute of Management
Noida, Uttar Pradesh, India

Amit Kumar Singh
Department of Commerce
Delhi School of Economics
New Delhi, India

Shuchita Singh
Professor
GL Bajaj Institute of Management and Research
Greater Noida, Uttar Pradesh, India

Jan Alexa Sotto
Cofounder Auspicia Co.
Manila, Philippines

Vijay Prakash Srivastava
Professor
Sri Dev Suman Uttarakhand University
Tehri Garhwal, Uttarakhand, India

Sitansu Ranjan Swain
PhD Scholar
SMS
GIET University
Gunupur, Odisha, India

D.D. Tewari
Professor
Department of Economics
Faculty of Commerce Administration & Law
University of Zululand
Richards Bay, South Africa

C.K. Tiwari
Associate Professor
Harcourt Butler Technical University
Kanpur, Uttar Pradesh, India

Jonah Tomno
Principal Officer
Star Discover Limited
Nairobi, Kenya

Introduction

Today, Africa is on the brink of a new era of major changes in the economy, summarized by such vectors as innovations, technologies, and the sustainability of business. Through these changes, women, the pillar of the African economy, are holding the central place in this revolution. Despite all these challenges, indigenes are steering change makers who are commercially minded, socially responsible, and feasible.

This book examines four main topics through which readers learn about the various impactful roles that women perform within African business and innovation systems.

The book analyzes African women's leadership in sustainable business practices through their sustainable entrepreneurship activities and their establishment of environmentally friendly business operations and social welfare services. It showcases women who use technology and engineering expertise to develop brand-new startup companies and groundbreaking engineering achievements for industry transformation.

Being a truly transformative leader entails that they don't rely mainly on their position of power, as attraction to change is intrinsic for a leader. Diverse roles in corporate settings, political roles, and entrepreneurial enterprises are allowing African women to set new leadership standards, creating new leadership roles.

The deepest understanding comes from real-life experiences of women with live cases. This book presents women's life stories and practical examples of those women who made it big, winning over obstacles, becoming influential in various professional domains.

These are testimonies of women who define Africa's future, who are resilient and capacitated with vision and impact. As readers analyze the valuable teachings of the described stories, they are encouraged to interact and unite to build an equitable, sustainable environment for everyone.

1 Diversity in Women's Entrepreneurship through Technological Advancement toward Sustainability in South Africa

A Study on Kenya

Sukanta Kumar Baral and D.D. Tewari

1.1 INTRODUCTION

Women's entrepreneurship in Africa has been on a remarkable growth trajectory in recent years, driven by the continent's economic transformation, technological advancements, and growing emphasis on gender equality. However, despite this progress, women entrepreneurs face unique challenges due to social, economic, and cultural factors that vary across different regions.

In Africa, women significantly contribute to the economy, often spearheading small- and medium-sized enterprises (SMEs), vital for economic growth, job creation, and poverty alleviation. According to the Global Entrepreneurship Monitor (GEM), Africa has the highest percentage of women entrepreneurs globally, with nearly 58% of African women self-employed. Yet their contributions to the formal economy are often underrepresented due to barriers such as limited access to funding, education, and business resources.

Several key factors have contributed to the rise of women entrepreneurs in Africa:

- *Economic Necessity:* Many African women turn to entrepreneurship to survive due to high unemployment rates and limited access to formal employment. This necessity-driven entrepreneurship often leads to the creation of informal businesses, particularly in sectors like retail, agriculture, and services.

DOI: 10.1201/9781003508953-1

- *Technological Advancements:* The proliferation of mobile phones and internet connectivity across Africa has opened new opportunities for women to engage in entrepreneurial activities. Mobile money platforms like M-Pesa in Kenya have revolutionized access to financial services, enabling women to run businesses remotely and access markets that were previously out of reach.
- *Government and International Initiatives:* Many African governments have recognized the importance of women's entrepreneurship in economic development and have introduced policies and programs to support women-owned businesses. International organizations, such as the World Bank and UN Women, have also implemented projects that provide training, mentorship, and funding to African women entrepreneurs.

Women entrepreneurs in Africa are primarily concentrated in the following sectors:

- *Agriculture:* Agriculture is the backbone of many African economies, and women play a key role in this sector, particularly in rural areas. Women entrepreneurs increasingly adopt sustainable farming practices, leveraging technology to increase productivity and reduce environmental impact.
- *Retail and Services:* Many women entrepreneurs operate small businesses in the retail and services sectors, such as clothing shops, food vendors, hairdressing salons, and cleaning services. These businesses are often informal and operate on a micro-scale, but they provide essential services to local communities.
- *Handicrafts and Artisanship:* Women also produce and sell traditional handicrafts, textiles, and handmade goods. This sector is a source of income and a way to preserve cultural heritage and traditions.
- *Tech Startups:* Although still underrepresented in the tech industry, there is a growing number of women-led tech startups in Africa. Women increasingly use digital platforms to launch businesses in e-commerce, fintech, health tech, and education tech.

Despite their entrepreneurial spirit, African women face numerous challenges in building and sustaining businesses:

- *Access to Finance:* African women entrepreneurs often struggle to secure venture funding. Traditional financial institutions are less likely to lend to women due to perceived risks and a lack of collateral. Moreover, women often have limited access to venture capital and angel investors, which is crucial for scaling businesses.
- *Cultural and Social Norms:* In many African societies, women are expected to prioritize family and domestic responsibilities over business activities. These societal expectations can limit women's ability to fully commit to their entrepreneurial ventures or scale them into larger operations.
- *Education and Skills Gap:* Many women entrepreneurs, especially in rural areas, lack formal education and business training, which hinders their

ability to manage and grow their enterprises effectively. A lack of access to training programs and professional networks exacerbates this skills gap.
- *Limited Market Access:* African women entrepreneurs often face barriers to entering formal markets. This is particularly true for women in rural areas, who may be isolated from larger markets due to poor infrastructure and a lack of digital literacy.

Despite these challenges, there is immense growth potential for women's entrepreneurship in Africa, particularly with the increasing adoption of technology and digital platforms. E-commerce, mobile banking, and social media have opened up new markets for women entrepreneurs, allowing them to reach consumers beyond their immediate geographic areas.

Additionally, sustainability and social impact are becoming central to many women-led businesses. Women entrepreneurs increasingly adopt environmentally friendly practices and align their businesses with the United Nations sustainable development goals (SDGs), particularly in areas like clean energy, responsible production, and inclusive economic growth.

Various organizations are working to support women's entrepreneurship in Africa. These include:

- *SheTrades (International Trade Centre):* This initiative connects women entrepreneurs to international markets by providing networking, training, and mentorship opportunities.
- *AWIEF (African Women Innovation & Entrepreneurship Forum):* AWIEF promotes innovation and entrepreneurship among African women through annual forums, training programs, and networking opportunities.
- *UN Women Empowerment Principles:* These principles guide businesses on empowering women in the workplace, marketplace, and community.

Women's entrepreneurship in Africa is vital to economic development and social change. Although African women entrepreneurs face numerous challenges, the increasing availability of technology, government support, and international initiatives present significant opportunities for growth. With continued investment in education, financial inclusion, and policy reform, African women entrepreneurs will be well-positioned to build sustainable and inclusive businesses contributing to the continent's development.

1.1.1 The Entrepreneurial Ecosystem in Kenya

The Entrepreneurial Ecosystem in Kenya has evolved into a dynamic space, contributing significantly to the country's economic growth. Diversity and sustainability have become essential components in driving inclusive growth and long-term success for businesses. This chapter explores the importance of these two aspects—diversity and sustainability—in Kenyan entrepreneurial ventures, supported by data and references from various sources.

1.1.2 Diversity in Entrepreneurial Ventures in Kenya

Diversity in entrepreneurship refers to including people from various backgrounds, such as gender, ethnicity, and geographic location. In Kenya, diversity has proven to drive innovation and market reach, enabling businesses to serve a wider array of consumers and adapt to evolving market conditions.

1.1.2.1 Gender Diversity

Kenya ranks among the top African countries in terms of female entrepreneurship. According to the Global Entrepreneurship Monitor (GEM) (2020), about 33% of early stage entrepreneurs in Kenya are women, compared to the global average of 23%. Women-owned businesses are particularly active in agriculture, retail, and services. However, the gap in funding and support remains an obstacle. A 2021 study by GIZ (Deutsche Gesellschaft für Internationale Zusammenarbeit) found that women in Kenya struggle with access to capital, business networks, and training compared to their male counterparts.

Economic Contribution: Women-led businesses contribute to about 20% of Kenya's GDP, according to the Kenya National Bureau of Statistics (KNBS). Enhancing gender diversity in entrepreneurship can increase Kenya's GDP by up to 12%, as estimated by the McKinsey Global Institute in their report "The Power of Parity" (2015). This underscores the economic importance of empowering women entrepreneurs through policies and programs focused on diversity and inclusion.

1.1.2.2 Ethnic and Cultural Diversity

Kenya's multiethnic society fosters various business ideas and products that cater to different cultural preferences. This diversity drives creativity and innovation in business ventures. A World Bank (2019) report on entrepreneurship in Kenya found that ethnically diverse businesses were likelier to innovate and introduce new products or services. Entrepreneurs from different ethnic backgrounds bring unique perspectives that can better address the needs of their communities.

Innovation and Market Reach: Ethnically diverse teams are more adaptable and responsive to market needs, particularly in sectors like food, fashion, and tourism, which are deeply influenced by cultural identities. Businesses that tap into Kenya's ethnic diversity can serve broader markets, both domestically and internationally, by tailoring their offerings to diverse consumer preferences.

1.1.2.3 Geographic and Economic Diversity

Kenya's economic landscape features urban hubs like Nairobi and Mombasa and rural areas where agriculture is dominant. Encouraging entrepreneurship in rural areas enhances economic inclusivity and reduces income inequality. According to the International Labour Organization (ILO), 80% of Kenya's rural population relies on agriculture, and entrepreneurial ventures in this sector can help bridge the gap between urban and rural economic opportunities.

Rural Entrepreneurship: A study by Brookings Institution (2021) revealed that promoting rural entrepreneurship in Kenya can lift millions out of poverty, given that 70% of Kenya's population lives in rural areas. Programs like the Kenya Women's Enterprise Fund target rural women entrepreneurs, fostering geographic diversity and economic sustainability.

1.2 SUSTAINABILITY IN ENTREPRENEURIAL VENTURES IN KENYA

Sustainability, in the context of entrepreneurship, refers to creating businesses that generate profit and contribute positively to environmental conservation and social well-being. Kenya's focus on sustainable entrepreneurship aligns with global trends and the SDGs. Entrepreneurs adopting sustainability practices in their ventures contribute to Kenya's long-term economic growth, reduce poverty, and address environmental challenges.

1.2.1 Environmental Sustainability

Kenya faces significant environmental challenges such as deforestation, climate change, and water scarcity, making sustainability a priority for many entrepreneurs. As a result, businesses that adopt eco-friendly practices and sustainable resource management are gaining competitive advantages.

Renewable Energy Sector: Kenya is a global leader in renewable energy, with 75% of its electricity generated from renewable sources, including geothermal, wind, and solar power, according to the International Energy Agency (IEA). Entrepreneurs in this sector are driving green innovations, creating jobs, and addressing climate change. For example, M-KOPA Solar, a Kenyan startup, provides solar-powered energy solutions to low-income households, helping over one million homes access clean energy while reducing carbon emissions.

Sustainable Agriculture: Entrepreneurs in this sector increasingly turn to organic farming, agroforestry, and water conservation techniques to enhance food security and protect ecosystems. These practices not only improve yields but also ensure long-term sustainability.

1.2.2 Social Sustainability

Social sustainability focuses on improving the well-being of communities through fair labor practices, education, and inclusive economic growth. Kenyan entrepreneurs who prioritize social impact contribute to the country's socioeconomic development.

Inclusive Employment and Fair Trade: Many Kenyan businesses are incorporating fair trade practices and employing marginalized communities, particularly women and youth. The *Kenya Fair Trade Organization (2020)* reported that entrepreneurs engaged in fair trade practices have improved the livelihoods of thousands of farmers and artisans, offering them better prices and working conditions. Social sustainability initiatives, such as the work of Amani ya Juu, which trains marginalized women to produce and sell handcrafted goods, demonstrate how entrepreneurial ventures can foster economic empowerment.

1.2.3 Economic Sustainability

Sustainable businesses are more resilient and better positioned for long-term success. By adopting sustainable practices, entrepreneurs can reduce operational costs, attract investment, and tap into growing consumer demand for eco-friendly and ethical products.

Green Jobs and Economic Resilience: According to the Kenya Private Sector Alliance (KEPSA), the shift toward green business models has created more than 300,000 jobs in sectors like renewable energy, waste management, and sustainable agriculture. Entrepreneurs who embrace sustainability can better mitigate risks such as resource scarcity, environmental regulations, and fluctuating market demand. In the 2020 report "Green Jobs in Kenya," KEPSA highlighted how sustainable enterprises were key to the country's post-COVID economic recovery, contributing to resilient supply chains and economic diversification.

1.2.3.1 The Intersection of Diversity and Sustainability

The combination of diversity and sustainability amplifies the potential for innovation and inclusive growth in Kenya's entrepreneurial ventures. Diverse entrepreneurs bring fresh perspectives that can solve complex sustainability challenges, while sustainable practices ensure long-term viability for businesses.

- *Case Study: Sanivation (Kenya):* Sanivation, a Kenyan startup, exemplifies the intersection of diversity and sustainability. Led by a diverse team, Sanivation focuses on transforming human waste into sustainable fuel, addressing both sanitation issues and environmental sustainability. The company's inclusive approach to hiring and training, especially in marginalized communities, enhances social equity, while its sustainable waste management practices reduce environmental impact.

1.2.3.2 Policy and Institutional Support

Government policies and institutional support are critical to promoting diversity and sustainability in Kenya's entrepreneurial landscape. Key initiatives include:

- *Kenya Vision 2030:* This national development blueprint emphasizes inclusive economic growth and sustainable development. Vision 2030's support for SMEs, particularly women and youth entrepreneurs, reflects Kenya's commitment to fostering diversity and sustainability.
- *Kenya Climate Innovation Center (KCIC):* Launched by the World Bank and the Kenyan government, KCIC provides funding, mentorship, and training to entrepreneurs focusing on sustainable business models. The center has supported over 300 startups in sectors like clean energy and agriculture, highlighting the importance of sustainability in entrepreneurship.

Diversity and sustainability are crucial drivers of entrepreneurial success in Kenya. Diverse entrepreneurial teams are better equipped to innovate and meet the needs of Kenya's multiethnic, geographically diverse population. At the same time, businesses that embrace sustainability contribute to environmental and social well-being and position themselves for long-term growth in an increasingly eco-conscious market. By fostering diversity and sustainability, Kenyan entrepreneurs can create resilient, inclusive, and successful ventures that drive national economic growth while addressing global challenges.

1.3 LITERATURE REVIEW

The following literature highlights the critical role of diversity in women's entrepreneurship through technological advancements and its contribution toward sustainability in South Africa and Kenya. While the intersection of these factors opens up opportunities for women to innovate and drive inclusive economic growth, structural barriers such as unequal access to finance and technology continue to pose challenges. Governments and stakeholders in both countries need to provide targeted support for women entrepreneurs, ensuring that technology and sustainability become integral to their business ventures.

Diversity in entrepreneurship is increasingly recognized as a vital element for innovation, growth, and social inclusion. Women's entrepreneurship, in particular, has emerged as a critical area of research, especially in Africa, where women play a significant role in economic development despite systemic barriers. Research by Minniti and Naudé (2010) argues that women's participation in entrepreneurial activities contributes to poverty alleviation, economic growth, and community development. Similarly, Chinomona and Maziriri (2015) highlight the potential of women entrepreneurs in African countries like South Africa and Kenya to foster inclusive economic participation, particularly in sectors like agriculture, manufacturing, and services.

In Kenya, women entrepreneurs constitute about *33% of all business owners*, as noted by the Global Entrepreneurship Monitor (GEM) (2020). However, they still face challenges related to access to finance, mentorship, and technological resources. Studies such as those by Kinyanjui (2006) emphasize the importance of tailored policy frameworks and financial institutions in reducing gender disparity in entrepreneurship.

The rapid technological advancements of the 21st century have created new opportunities for women entrepreneurs. Digital tools such as mobile banking, e-commerce platforms, and digital marketing have enhanced access to markets and resources, especially for women-led businesses in developing countries. In a study by Kariuki and Ndunge (2018), technology is identified as a significant enabler for women entrepreneurs in Kenya, providing tools for marketing, sales, and business management that were previously inaccessible.

In South Africa, Harvey and Griffith (2007) discuss the role of information and communication technology (ICT) in fostering female entrepreneurship, especially in rural areas. Mobile phones, online platforms, and social media have opened new opportunities for women entrepreneurs to reach a broader customer base, allowing them to engage in trade and services previously limited by geographic and logistical constraints. Chisasa (2014) similarly notes that mobile money platforms, such as M-Pesa in Kenya, have revolutionized the ability of women to access financial services, particularly in rural and underserved communities.

Sustainability is becoming essential in business ventures across Africa, driven by environmental concerns and the need for long-term economic resilience. Women entrepreneurs are increasingly playing a role in sustainable business practices, particularly in agriculture and renewable energy sectors. According to Nziku and Struthers (2019), women entrepreneurs in Kenya and South Africa are pivotal in driving sustainable business practices, often focusing on eco-friendly solutions and social enterprise models that benefit the environment and their communities.

Research by Maina and Muturi emphasizes that sustainability-driven women entrepreneurs in Kenya are heavily involved in sectors like agriculture, particularly in promoting organic farming and agroforestry. These ventures not only improve economic outcomes for women but also contribute to food security and environmental sustainability. Similarly, Kariuki and Mulei (2020) argue that technology can enhance sustainability practices among women entrepreneurs by enabling efficient resource management and innovative production techniques.

The intersection of diversity, technological advancement, and sustainability is particularly important for women entrepreneurs in Africa. Studies such as those by Simpson et al. (2017) highlight that technological tools provide women entrepreneurs with access to new markets, improve their business operations, and help them incorporate sustainable practices. In South Africa, Fatoki (2014) stresses that technology and innovation enable women entrepreneurs to navigate the challenges posed by market competition and environmental sustainability.

Moreover, in Kenya, using digital tools such as mobile payment platforms and e-commerce has improved the operational efficiency and profitability of women-led enterprises, allowing them to scale their businesses while incorporating eco-friendly practices. Makena et al. (2014) argue that technology enables women entrepreneurs to manage resources better and engage in sustainable production methods, particularly in the agriculture and clean energy sectors.

Despite the positive strides in women's entrepreneurship, there remain significant challenges. Amorós and Bosma (2014) indicate that access to finance, education, and business networks continues to be a barrier for women entrepreneurs, especially in sub-Saharan Africa. This is compounded by cultural and structural biases that limit women's ability to leverage technological advancements in their businesses fully. In South Africa and Kenya, gender inequality in access to financial capital and technology creates hurdles for women entrepreneurs, as documented by Meunier et al. (2019).

In addition, the sociocultural factors in both countries often hinder women's access to resources that could enhance their business sustainability. Kantor notes that gender roles and expectations can limit women's time and mobility, making it difficult for them to engage fully in entrepreneurial activities or take advantage of technological tools. Furthermore, as Kariuki and Ndunge (2018) argue, the lack of adequate digital literacy and training for women in rural areas exacerbates these challenges.

1.4 OBJECTIVES

1. *To Examine the Role of Diversity in Women's Entrepreneurship:* The first objective is to explore how diversity, including cultural, socioeconomic, and educational factors, influences the entrepreneurial activities of women in South Africa and Kenya.
2. *To Investigate the Impact of Technological Advancement on Women Entrepreneurs:*
   This objective aims to assess the role of technological advancements, such as mobile banking, e-commerce platforms, and digital tools, in empowering women entrepreneurs.
3. *To Explore the Relationship between Women's Entrepreneurship and Sustainability:*
   The research will focus on how women entrepreneurs integrate sustainability into their business models, particularly in agriculture, manufacturing, and energy sectors.

1.5 ANALYSIS AND DISCUSSION

1.5.1 Objective-1

The first objective seeks to understand how multidimensional diversity—encompassing cultural, socioeconomic, and educational aspects—affects women's entrepreneurial activities in South Africa and Kenya. This detailed analysis highlights the impact of these diversities on innovation, market expansion, and business growth among women entrepreneurs in both countries.

Cultural Diversity and Women's Entrepreneurship: Cultural diversity plays a crucial role in shaping women's entrepreneurial mindset and approach. South Africa and Kenya are among Africa's most culturally diverse countries, with each country having multiple ethnicities, languages, and traditions, all of which have varying impacts on women entrepreneurs.

Cultural Norms and Business Practices: In South Africa, the country's post-apartheid socioeconomic landscape has seen significant efforts to integrate diverse communities, including previously marginalized women. However, cultural factors can either empower or restrict women's entrepreneurial activities. Women from different ethnic communities, such as the Zulu, Xhosa, and Sotho, often exhibit unique business practices shaped by local traditions and values. For example, in rural areas, culturally rooted gender roles might limit women's access to certain types of businesses. In contrast, in urban areas, the multicultural environment often encourages more diverse business ventures.

In Kenya, cultural norms also significantly influence entrepreneurial activities. Among ethnic groups like the Kikuyu, Luo, and Kalenjin, there are different societal expectations for women, which affect how they participate in entrepreneurship. A 2019 study by Kariuki and Wanjiku showed that Kikuyu women, for instance, are more likely to engage in trade and entrepreneurial activities due to historical roles in commerce, while women in more pastoralist communities face restrictions that limit business participation. The cultural acceptance of women as business owners can either enhance or limit their opportunities, thus directly affecting the diversity of businesses that women engage in.

Innovation through Cultural Knowledge: Cultural diversity fosters innovation by allowing women to incorporate traditional practices with modern business strategies. In both South Africa and Kenya, women entrepreneurs, especially in the agricultural and handicraft sectors, often leverage indigenous knowledge systems to create unique products that resonate with both local and international markets. For example, South African women working in the organic skincare industry utilize locally sourced natural ingredients, blending traditional healing methods with modern cosmetic production techniques. Similarly, Kenyan women entrepreneurs involved in handwoven textile production integrate age-old weaving practices, appealing to global markets that appreciate authentic, handcrafted goods. According to a World Bank report (2020), South African women-led businesses in cultural sectors such as arts, crafts, and ecotourism have grown by 18% annually, driven by innovations that merge traditional and contemporary knowledge systems.

1.5.1.1 Socioeconomic Diversity and Women's Entrepreneurship

Socioeconomic diversity refers to the differences in income levels, access to resources, and economic mobility that shape women's entrepreneurial activities in South Africa and Kenya. These differences are critical in influencing the type of ventures women can engage in, their growth potential, and their long-term sustainability.

Access to Financial Capital: In both South Africa and Kenya, socio-economic disparities among women entrepreneurs significantly impact their ability to access financial resources. Women from wealthier backgrounds are more likely to access formal financing options such as bank loans and venture capital, enabling them to scale their businesses more easily. In contrast, women from lower-income communities often rely on informal financing options, such as savings groups, family contributions, and microfinance institutions. For example, Kenya's Women's Enterprise Fund has been crucial in providing loans to women entrepreneurs in underserved communities, with over 60% of its beneficiaries coming from low-income rural areas.

In South Africa, the Broad-Based Black Economic Empowerment (BBBEE) policy has helped to increase financial access for black women entrepreneurs. However, there are still socio-economic gaps in the entrepreneurial ecosystem, with 60% of South African women-owned businesses citing lack of access to credit as a major barrier to growth (GEM, 2020).

Income Disparities and Business Growth: Income levels also influence women-led businesses' scale and growth potential. Women from higher socio-economic backgrounds tend to run businesses in high-growth sectors such as technology, real estate, and manufacturing. In contrast, women from lower-income backgrounds often engage in small-scale retail, agriculture, and home-based businesses. This creates a diverse entrepreneurial landscape where women at different economic levels contribute to various sectors of the economy. According to the Global Entrepreneurship Monitor (2020), 45% of women entrepreneurs in Kenya are engaged in small-scale businesses, while 18% are in medium-sized enterprises. In South Africa, 52% of women entrepreneurs run small enterprises. Still, those in higher-income groups are more likely to scale their businesses to medium or large ventures, with 25% of women entrepreneurs in higher-growth industries.

1.5.1.2 Educational Diversity and Women's Entrepreneurship

Educational diversity refers to the varying levels of formal and informal education that women entrepreneurs possess, significantly affecting their business activities, innovation capacity, and ability to adopt new technologies.

Formal Education and Business Acumen: Women with higher levels of formal education tend to be more adept at leveraging complex business models, understanding market trends, and utilizing advanced technologies. For example, women entrepreneurs with university degrees in Kenya are more likely to engage in tech-based startups, financial services, or export-oriented businesses. These women have the skills to navigate regulatory frameworks, access digital platforms, and adopt modern business practices, giving them a competitive edge in local and international markets.

In South Africa, government and private initiatives like the South African Women Entrepreneurs' Network (SAWEN) have provided educational programs to enhance women's business skills, particularly in sectors like manufacturing and green energy. Women entrepreneurs with a solid educational background are more likely to engage in high-growth sectors such as technology, green industries, and professional services.

Informal Education and Skill Development: Women with lower levels of formal education often rely on informal learning, such as vocational training, apprenticeship, or hands-on experience. In rural areas of Kenya, for instance, women in the agriculture sector typically acquire skills through generational knowledge transfer. These women may have limited formal education but excel in businesses that require specialized, practical skills. Similarly, in South Africa, women in sectors like crafts and tourism often operate successful ventures by leveraging traditional knowledge combined with training programs provided by NGOs or community groups. The ILO (2018) reported that 67% of women entrepreneurs in rural Kenya rely on informal education or skill acquisition to operate businesses, while in South Africa, 58% of women entrepreneurs in rural areas receive business support from local training programs.

1.5.1.3 Diversity as a Driver of Innovation, Market Expansion, and Business Growth

Innovation through Diverse Perspectives: Cultural and socio-economic diversity foster innovation among women entrepreneurs by encouraging the blending of different perspectives, skills, and ideas. For instance, in South Africa, women entrepreneurs from different cultural backgrounds bring varying customer insights and market needs into their businesses. This leads to product design, service delivery, and business model innovation. For example, in Kenya's fintech sector, women entrepreneurs have driven the development of financial solutions tailored to underserved populations, particularly rural women.

Market Expansion and New Opportunities: Diversity enables women entrepreneurs to access a wider range of markets. Entrepreneurs who understand and incorporate cultural diversity into their products and services can cater to both local and global markets. For example, Kenyan women in the textile and crafts industries export products based on traditional designs to international markets, capitalizing on the global demand for authentic, handcrafted goods. A report in 2019 by UN Women found that women-led businesses in Kenya's handicraft sector grew by 35% in export revenue between 2016 and 2019, largely due to innovations in product design and marketing strategies that appealed to diverse markets.

Business Growth and Economic Impact: The diversity in women's entrepreneurship contributes to broader economic growth in South Africa and Kenya. Women entrepreneurs contribute significantly to agriculture, tourism, crafts, and services, all essential for economic development. By fostering an inclusive environment that supports diverse women entrepreneurs, both countries can enhance job creation, reduce poverty, and promote economic resilience. According to the World Bank (2020), women-owned businesses in South Africa and Kenya contribute 13% of their respective GDPs and employ over 30% of the workforce in agriculture and services.

1.5.2 Objective-2

Diversity in women's entrepreneurship in South Africa and Kenya—whether cultural, socio-economic, or educational—plays a vital role in shaping business practices, innovation, and market expansion. Women entrepreneurs from diverse backgrounds bring unique perspectives and skills that drive business growth and contribute to the overall economic development of both nations. By examining how diversity influences these factors, this research aims to provide valuable insights into creating more inclusive, innovative, and sustainable entrepreneurial ecosystems in Africa.

The study seeks to comprehensively evaluate the transformative role of technological advancements in empowering women entrepreneurs, with a particular focus on how innovations such as mobile banking, e-commerce platforms, and digital tools are shaping the entrepreneurial landscape in South Africa and Kenya. In recent years, the rapid pace of technological change has presented new opportunities for women entrepreneurs to overcome traditional barriers, expand their markets, and enhance business efficiency. This analysis will explore the multifaceted impact of these technologies on the financial inclusion, market access, operational efficiency, and overall business growth of women-owned enterprises.

1.5.2.1 Mobile Banking and Financial Inclusion

Mobile banking has emerged as a crucial tool in enhancing financial inclusion for women entrepreneurs, particularly those in rural and underserved areas. In many parts of South Africa and Kenya, women have historically faced challenges in accessing formal banking services due to cultural norms, lack of documentation, or limited financial literacy. Mobile banking platforms have played a pivotal role in overcoming these barriers.

Financial Access and Capital Mobilization: One of the key impacts of mobile banking is its ability to provide women entrepreneurs with access to financial services that were previously out of reach. In Kenya, platforms like M-Pesa, launched by Safaricom, have revolutionized the way women manage and access their finances. M-Pesa allows women to open mobile wallets, save money, transfer funds, and receive payments, all without the need for a traditional bank account. This has been particularly beneficial for women running small businesses, enabling them to receive payments directly from customers and access microloans to fund their ventures.

In South Africa, mobile banking platforms such as FNB's eWallet have enabled women in urban and rural areas to manage their businesses more efficiently. These platforms are especially useful for entrepreneurs in agriculture, retail, and informal trading, where cash flow management and access to working capital are critical. According to the Central Bank of Kenya report in 2019, 86% of women entrepreneurs in rural areas use mobile banking services like M-Pesa to manage their businesses, while 52% of women-owned SMEs in South Africa use mobile banking for financial transactions.

Empowerment through Financial Independence: Mobile banking has empowered women entrepreneurs by providing them with a level of financial independence that was previously unavailable. It has also reduced their reliance on intermediaries, making it easier for them to control their finances and reinvest in their businesses. The ability to save, borrow, and transact independently has given many women entrepreneurs the confidence to grow their businesses and take on more significant ventures.

1.5.2.2 E-Commerce Platforms and Market Access

E-commerce platforms have drastically transformed how women entrepreneurs conduct business, enabling them to reach wider markets locally and globally. The rise of digital marketplaces and online sales channels has eliminated many geographical and infrastructural barriers that previously limited the growth of women-owned businesses.

Expansion of Customer Base: In Kenya, e-commerce platforms such as Jumia and Twiga Foods have enabled women entrepreneurs, particularly in the retail and agricultural sectors, to reach a larger customer base beyond their immediate communities. Women who sell handmade crafts, clothing, and agricultural produce can now list their goods on these platforms, allowing them to tap into urban markets and even international buyers. This digital shift has been particularly significant for rural women, who have traditionally had limited access to urban marketplaces due to poor infrastructure or high transportation costs.

In South Africa, platforms like Takealot and Bidorbuy have created similar opportunities for women entrepreneurs to expand their businesses. Through these platforms, women selling products ranging from home goods to beauty products can access a diverse customer base, increasing their sales volumes and business visibility. Moreover, the ease of setting up an online store has democratized the e-commerce space, enabling women with limited capital or physical storefronts to participate in the market. A study by UNCTAD showed that 41% of women entrepreneurs in Kenya use e-commerce platforms to sell their products, and these businesses reported an average revenue increase of 25% after transitioning to online sales. In South Africa, 35% of women entrepreneurs reported using e-commerce platforms, with 67% stating that it helped them access new markets.

Reduced Barriers to Entry: E-commerce platforms have also reduced the barriers to entry for women entrepreneurs by eliminating the need for large physical stores, reducing overhead costs, and providing access to a wide array of digital marketing tools. Many women entrepreneurs in South Africa and Kenya have utilized these platforms to sell niche products, such as handmade goods, organic skincare products, and local fashion, which cater to specific customer segments that may not be accessible through traditional retail channels.

1.5.2.3 Digital Tools and Operational Efficiency

Adopting digital tools, including business management software, social media marketing platforms, and digital payment systems, has significantly improved the operational efficiency of women entrepreneurs in South Africa and Kenya. These tools allow entrepreneurs to streamline their operations, enhance customer engagement, and optimize resource management.

Business Management and Automation: Digital tools such as accounting software, inventory management systems, and customer relationship management (CRM) platforms have simplified business operations for women entrepreneurs. For instance, in Kenya, women running SMEs in sectors like retail and hospitality are using digital tools to automate tasks such as invoicing, inventory tracking, and payroll management. This automation allows entrepreneurs to focus more on growth and innovation rather than spending excessive time on administrative tasks.

In South Africa, women entrepreneurs are increasingly adopting cloud-based tools that enable them to manage their businesses remotely. For example, women in the tourism industry use digital booking and payment systems to handle reservations and customer interactions, significantly improving their ability to manage their businesses in real time. A survey by the International Finance Corporation (IFC) in 2018 found that women entrepreneurs who use digital tools for business management reported a 32% improvement in operational efficiency and a 15% reduction in operational costs. Similarly, 64% of women-owned SMEs in Kenya that adopted digital tools reported improved business performance.

Social Media Marketing and Customer Engagement: Social media platforms such as Facebook, Instagram, and WhatsApp have become essential marketing tools for women entrepreneurs. These platforms allow them to engage directly with customers, build brand awareness, and promote their products and services relatively cheaply. In South Africa and Kenya, women entrepreneurs have successfully used social media to reach wider audiences, launch new products, and interact with customers in real time.

For instance, women in Kenya's fashion and beauty sectors have leveraged Instagram to showcase their products, attract new customers, and build brand loyalty. In South Africa, women in the food and beverage industry frequently use Facebook and WhatsApp to promote their products and coordinate deliveries with customers, particularly during social restrictions, such as the COVID-19 pandemic. According to a 2021 study by Hootsuite, 58% of women entrepreneurs in South Africa reported that social media marketing was critical to their business success, with 45% stating that it directly led to increased sales. In Kenya, 73% of women-owned businesses used social media for marketing, with the given 68% attributing business growth to social media engagement.

1.5.2.4 Technology as a Driver of Business Growth

Technological advancements have improved operational efficiency and enabled women entrepreneurs to scale their businesses and explore new growth opportunities. Integrating mobile banking, e-commerce platforms, and digital tools has helped women-led businesses move from local to national and even international markets.

Scaling Businesses through Technology: Women entrepreneurs who adopt technology early often experience faster business growth. For example, in South Africa, women in the renewable energy and technology sectors use cutting-edge innovations to solve local problems, such as providing solar energy solutions to rural communities. In Kenya, women in the agri-tech sector use mobile applications to connect farmers with markets, improve crop yields, and provide real-time data on market prices, enabling their businesses to scale and reach new levels of success.

Improving Competitiveness and Resilience: Technology has also improved the competitiveness and resilience of women-owned businesses, particularly in times of crisis. For instance, during the COVID-19 pandemic, women entrepreneurs in both South Africa and Kenya who had already adopted digital tools could continue operations through online platforms. At the same time, those without access to technology faced greater disruptions. The ability to pivot to online sales and digital operations has made technology an essential factor in ensuring women-owned enterprises' long-term survival and growth. A World Economic Forum report in 2020 revealed that women-led businesses in Kenya that adopted e-commerce early on experienced a 40% growth rate, compared to 17% for those that relied solely on physical storefronts. In South Africa, women in the tech sector saw a 50% increase in demand for digital solutions during the pandemic, driving further business growth.

Technological advancements, including mobile banking, e-commerce platforms, and digital tools, have had a transformative impact on women entrepreneurs in South Africa and Kenya. These technologies have empowered women by improving financial inclusion, expanding market access, enhancing operational efficiency, and driving business growth. By adopting and leveraging these technologies, women entrepreneurs are better positioned to overcome traditional barriers, scale their businesses, and contribute to the broader economic development of their regions.

1.5.3 Objective-3

Further, the study aims to investigate the intricate relationship between women's entrepreneurship and sustainability, focusing on how women entrepreneurs in South Africa and Kenya actively integrate sustainable practices into their business models. The research will specifically target three key sectors: agriculture, manufacturing, and energy. By analyzing these sectors, the study will reveal how women entrepreneurs contribute to sustainable development while simultaneously achieving economic growth and resilience. This exploration will encompass the challenges they face, their innovative strategies, and their sustainable practices' impact on both their businesses and the broader community.

1.5.3.1 Integration of Sustainable Practices in Agriculture

The agricultural sector is a vital area where women entrepreneurs can implement sustainable practices that benefit their businesses and the environment. In both South

Africa and Kenya, women play a crucial role in agriculture, and their integration of sustainability is increasingly important for food security, environmental conservation, and economic empowerment.

Sustainable Farming Techniques: Women entrepreneurs in agriculture are adopting sustainable farming techniques that promote environmental stewardship, such as organic farming, agroecology, and permaculture. These methods help to preserve soil health, reduce chemical usage, and enhance biodiversity. For instance, in Kenya, many women farmers are transitioning from traditional farming practices to organic methods, improving crop yields and creating products in high demand in both local and international markets.

Climate Resilience: Women-led agricultural enterprises are also focusing on building climate resilience. They implement practices such as crop rotation, intercropping, and water conservation techniques to adapt to changing climate conditions. In South Africa, women farmers in areas affected by drought use rainwater harvesting systems and drought-resistant crop varieties to mitigate the impacts of climate change. These sustainable practices enhance their productivity and contribute to the broader goals of environmental sustainability and community resilience. A study by the Food and Agriculture Organization of the UN (2021) found that women-led farms employing sustainable practices in Kenya reported a 30% increase in yield compared to conventional farms, highlighting the economic viability of sustainable agriculture.

Community Engagement and Education: Women entrepreneurs are often at the forefront of community engagement and education regarding sustainable practices. They serve as role models and educators within their communities, sharing knowledge about sustainable farming, nutrition, and environmental conservation. This enhances their business prospects and fosters a culture of sustainability within their communities.

1.5.3.2 Sustainable Manufacturing Practices

In the manufacturing sector, women entrepreneurs are increasingly integrating sustainability into their business operations, focusing on reducing waste, utilizing eco-friendly materials, and promoting ethical production processes.

Eco-Friendly Products: Women-led manufacturing businesses are innovating to produce eco-friendly products that cater to a growing market demand for sustainability. For example, in South Africa, women entrepreneurs in the textile industry use organic cotton and recycled materials to create sustainable fashion lines. Similarly, in Kenya, women in the handicraft sector utilize locally sourced materials and traditional craftsmanship techniques to produce sustainable and culturally significant goods.

Waste Reduction and Circular Economy: Many women entrepreneurs are adopting waste reduction strategies and principles of the circular economy, where waste is minimized, and materials are reused and recycled. Women-led enterprises are exploring innovative ways to repurpose waste materials, such as turning plastic waste into building materials or crafting artisanal products. These initiatives reduce environmental impact and create new economic opportunities and jobs within their communities. According to a report by UN Women, women-owned manufacturing businesses that implemented sustainable practices in South Africa experienced a 20% increase in customer loyalty and brand recognition, showcasing the market value of sustainability.

Sustainable Supply Chains: Women entrepreneurs also create sustainable supply chains prioritizing ethical sourcing, fair labor practices, and environmental responsibility. By establishing partnerships with local suppliers and ensuring that their sourcing practices align with sustainability goals, women entrepreneurs contribute to the overall sustainability of their industries.

1.5.3.3 Sustainability in the Energy Sector

The energy sector is another critical area where women entrepreneurs can make significant contributions to sustainability, particularly in the context of renewable energy solutions.

Promotion of Renewable Energy: Women entrepreneurs are leading initiatives to promote renewable energy solutions such as solar, wind, and bioenergy. In Kenya, for example, women-led enterprises are installing solar panels in rural areas, providing access to clean energy and creating jobs in installation and maintenance. These initiatives enhance energy access, reduce greenhouse gas emissions, and promote environmental sustainability.

Energy Efficiency Practices: Women-owned businesses in the energy sector also focus on energy efficiency practices that reduce consumption and operational costs. By adopting energy-efficient technologies and practices, such as using LED lighting and energy-efficient appliances, women entrepreneurs can lower their operational costs while contributing to energy conservation. A report by IRENA indicated that women-led renewable energy projects in Kenya can potentially increase energy access by 25% in underserved communities, illustrating the impact of women's entrepreneurship on sustainable energy solutions.

Community Empowerment through Energy Access: Women entrepreneurs in the energy sector often play a crucial role in empowering their communities by providing access to clean energy. This access improves living conditions and enables economic activities such as education, healthcare, and small business development. Women-led energy initiatives contribute to social and economic empowerment, fostering sustainable development at the community level.

1.5.3.4 Challenges and Barriers to Sustainability Integration

Despite the positive impacts of women's entrepreneurship on sustainability, challenges and barriers still exist that hinder the full integration of sustainable practices in their businesses.

Access to Resources and Training: Women entrepreneurs often face barriers in accessing the resources and training necessary for implementing sustainable practices. Limited access to finance, lack of technical training, and insufficient information about sustainable technologies can restrict their ability to adopt eco-friendly practices.

Cultural and Societal Norms: Cultural and societal norms can also pose challenges, particularly in more traditional settings where women's roles in business may be undervalued. Addressing these norms and advocating for women's rights in entrepreneurship is essential for promoting sustainability in their ventures.

Policy and Institutional Support: The lack of supportive policies and institutional frameworks can hinder women's ability to integrate sustainability into their businesses. Effective government policies that promote gender equality, access to finance, and sustainability can enhance the role of women entrepreneurs in driving sustainable development.

Exploring the relationship between women's entrepreneurship and sustainability reveals women's significant contributions in integrating sustainable practices into various sectors, particularly agriculture, manufacturing, and energy. By leveraging sustainable business models, women entrepreneurs enhance their economic prospects and contribute to environmental conservation and community resilience. This research will provide insights into the innovative strategies women employ, the challenges they face, and the broader implications for sustainable development in South Africa and Kenya. Understanding this relationship is vital for promoting gender-inclusive policies and practices that foster a sustainable future.

1.6 CONCLUSION

The exploration of diversity in women's entrepreneurship through technological advancement towards sustainability in South Africa, with a special reference to Kenya, underscores the critical role that women entrepreneurs play in driving economic growth, fostering innovation, and promoting sustainable practices. This research highlights how women entrepreneurs, fueled by technological advancements, leverage diversity in cultural, socio-economic, and educational backgrounds to create resilient businesses that contribute to their communities and address pressing environmental challenges. Integrating mobile banking, e-commerce platforms, and digital tools has significantly empowered women entrepreneurs to overcome traditional barriers and expand their market reach. In South Africa and Kenya, women harness these technologies to streamline operations, enhance customer engagement, and foster sustainable practices across various sectors, including agriculture, manufacturing, and energy. The findings indicate that women-led enterprises that embrace sustainability achieve economic success and contribute to the broader goals of environmental conservation and community well-being. The interconnectedness of diversity and sustainability in entrepreneurship is found globally. Women entrepreneurs from diverse backgrounds bring unique perspectives and solutions, driving innovation and creativity in their business models. This diversity enriches the entrepreneurial ecosystem and enhances the potential for sustainable development by addressing the needs of different communities and markets. However, the chapter highlights women entrepreneurs' challenges, including limited resource access, training, and supportive policies. To fully realize the potential of women's entrepreneurship in driving sustainability, governments, financial institutions, and civil society need to collaborate in creating an enabling environment that supports women entrepreneurs. This includes promoting gender equality, providing access to finance, and offering training programs that equip women with the skills needed to thrive in an increasingly digital and sustainable economy. In conclusion, fostering diversity in women's entrepreneurship through technological advancement is a pathway to economic empowerment for women in South Africa and Kenya and a crucial strategy for achieving sustainable development goals. As women continue to lead the charge in integrating sustainability into their business practices, their contributions will play a pivotal role in shaping a more inclusive, equitable, and sustainable future for all. Embracing and supporting these efforts will ultimately benefit society, creating a robust entrepreneurial landscape that champions diversity, innovation, and sustainability.

REFERENCES

Amorós, J. E., & Bosma, N. (2014). *Global entrepreneurship monitor 2013 global report*. Global Entrepreneurship Research Association.

Brookings Institution. (2021). *African entrepreneurs: Key to lifting people out of poverty*. Brookings. Retrieved from https://www.brookings.edu/research/african-entrepreneurs-key-to-lifting-people-out-of-poverty/

Chinomona, E., & Maziriri, E. T. (2015). Women in action: Challenges facing women entrepreneurs in the Gauteng province of South Africa. *International Business & Economics Research Journal (IBER)*, *14*(6), 835–850.

Chisasa, J. (2014). Financial inclusion through mobile money: The case of women entrepreneurs in Kenya. *International Journal of Business and Economic Research*, *6*(5), 75–89.

Fatoki, O. (2014). The entrepreneurial orientation of micro enterprises in the retail sector in South Africa. *Mediterranean Journal of Social Sciences*, *5*(16), 121–129.

Global Entrepreneurship Monitor. (2020). *GEM 2020–2021 global report*. Global Entrepreneurship Research Association. Retrieved from https://www.gemconsortium.org/report/gem-2020–2021-global-report

Harvey, M., & Griffith, D. (2007). Women's entrepreneurship in sub-Saharan Africa: Empirical evidence from the informal sector in Kenya. *African Development Review*, *19*(2), 253–277.

International Labour Organization (ILO). (2018). *ILO report highlights importance of rural entrepreneurship in reducing poverty*. ILO. Retrieved from https://www.ilo.org/global/about-the-ilo/newsroom/news/WCMS_731060/lang-en/index.htm

Kariuki, M., & Mulei, K. (2020). Technology and sustainability practices among women entrepreneurs in Kenya. *Journal of Sustainable Entrepreneurship*, *4*(2), 13–22.

Kariuki, N., & Ndunge, P. (2018). ICT and women entrepreneurship in Kenya. *Journal of African Business*, *9*(1), 45–63.

Kenya National Bureau of Statistics (KNBS). (2020). *Kenya's economic survey report*. KNBS. Retrieved from https://www.knbs.or.ke/

Kenya Private Sector Alliance (KEPSA). (2020). *Green jobs in Kenya: Creating jobs and building resilience through green business models*. KEPSA. Retrieved from https://www.kepsa.or.ke/

Kinyanjui, M. N. (2006). Overcoming barriers to enterprise growth: The experience of women-owned firms in Nairobi. *International Journal of Gender and Entrepreneurship*, *3*(2), 69–86.

Makena, J., Kubaison, S. T., & Njati, I. C. (2014). Challenges facing women entrepreneurs in accessing business finance in Kenya: Case of Ruiru Township, Kiambu County. *Journal of Business and Management*, *16*(4), 83–91.

McKinsey Global Institute. (2015). *The power of parity: How advancing women's equality can add $12 trillion to global growth*. McKinsey & Company. Retrieved from https://www.mckinsey.com/featured-insights/employment-and-growth/how-advancing-womens-equality-can-add-12-trillion-to-global-growth

Meunier, F., Krylova, Y., & Ramalho, R. (2019). *Women entrepreneurship: How to measure the gap between new female and male entrepreneurs?* World Bank Group.

Minniti, M., & Naudé, W. (2010). What do we know about the patterns and determinants of female entrepreneurship across countries? *European Journal of Development Research*, *22*(3), 277–293.

Nziku, D. M., & Struthers, J. (2019). Female entrepreneurship and sustainability in sub-Saharan Africa. *Journal of Global Business and Technology*, *15*(1), 23–36.

World Bank. (2019). *Doing business in Kenya: Economic profile and entrepreneurship*. World Bank Group. Retrieved from https://www.worldbank.org/en/country/kenya/publication/doing-business-2019

2 Empowering Change

The Role of Diverse Women Entrepreneurs in Shaping Africa's Sustainable Future

Supriya Lamba Sahdev, Chitra Krishnan, Gary W. Paul, and Rashika Panday

2.1 INTRODUCTION

During the past few decades, Africa has apparently been through actual structural changes, and economic diversification is greatly characterized by entrepreneurship. The centrality of women entrepreneurs has become a major force for development, innovation, and social transformation. For a while now, African women entrepreneurs have defied conventional social norms, engaged in business, and contributed to their families and communities and in so doing transformed the economy. Overall, women in many African countries are visible especially in areas that have for long been dominated by men in industries like manufacturing, technology, and agriculture and energy. GEM, which is The Global Entrepreneurship Monitor that is a globally recognized annual research, shows that sub-Saharan Africa has the highest female absorption rate in entrepreneurship. For instance, over 50% of the business owners in both Ghana and Uganda are women, as compared to men (Falaiye et al., 2024). It signifies gradual, yet evident, progressive changes in gender relations, where women are not restricted to performing mere subsistence economic activities but are leading other economic responsibilities in both formal and informal sectors. Nevertheless, the expansion of wide local excision in Africa faces challenges as discussed later. There are restrictions such as access to finance, legal and cultural barriers, and education and networks of business production, which are barriers African women entrepreneurs encounter. The fact that women and men do not access resources and opportunities in the same ways has resulted in what the literature refers to as the "gender gap" in entrepreneurship whereby women are constrained to engaging in informal finance and operate in small businesses (Fatimah et al., 2024). Despite these challenges, which have limited the growth of women entrepreneurs, most of them have adopted measures for coping and creating new ways of combating these challenges through networking, use of technology, and social entrepreneurship. To begin with, one major factor has been the increasing focus on women's economic empowerment in Africa. The following are some of the government and nongovernmental organizations' attempts to assist women entrepreneurs: (Sanchez-Riorifo et al., 2024).

 DOI: 10.1201/9781003508953-2

Such initiatives include capacity development programs, micro financing, coaching, and legislation by governments that seek to address women's marginalization in business. Nevertheless, the indicated problems remain constant, yet African women entrepreneurs are gradually acknowledged as promoters of diversified and sustainable economy. Gender diversity is essential when it comes to different aspects of entrepreneurship, especially since the future belongs to more diverse, more resistant, and more sustainable innovations (Nayak & Hegde, 2023). The entrepreneurship diversity in the ecosystems offers multiple prospective skills sets and experiences required to solve various societal issues like poverty, inequality, and climate change. Women entrepreneurs usually have different approach, strengths, and values, especially those from African countries, when they address social issues and sustainable development. Across many of the African societies, the women have always been charged with the responsibility tasks of resource management within households, communities, and even farms. They are, therefore, able to rely on this experience to develop new ideas of how people should manage resources and develop sustainable and viable structures and how communities should be developed by coming up with different models. Women's ideas and perspectives in entrepreneurship are basic and important for the economic development of societies to be sustainable and inclusive. In addition, participation in entrepreneurship has been found to promote social inclusion especially for those who have been locked out of the mainstream economy like women, youths, and those from rural settings. If more women entrepreneurs are promoted and supported across African countries, more talent and solutions can be drawn from such a diverse population for improved economic transformation in the continent. The involvement of multiple viewpoints based on gender, region, or social experience contributes positively to the business environment and encourages innovation and higher quality of solutions and outcomes (McAdam, 2013). The relationship between female entrepreneurial activities and global sustainability—female entrepreneurship represents an essential aspect in the overall drive toward sustainability—now being viewed as having a significant importance toward the achievement of most of the sustainable development goals (SDGs) established by the United Nations. The SDGs were agreed in 2015 and are a set of goals that the global community wants to achieve to help address some of the world's most pressing issues such as poverty, inequality, and climate change. Women's entrepreneurship is connected to many of these goals, especially to Goal 5 on gender equality, Goal 8 on decent work and economic growth, and Goal 12 on responsible consumption and production.

Women entrepreneurs in Africa are contributing immensely toward enhancing these objectives by establishing enterprises that operate on profit with an equal balance on social and ecological responsibilities. Most women's businesses operate in the informal segment, targeting industries like farming and farming-related businesses, teaching and nursing, and clean energy, the critical aspects of sustainable development (Odeyemi et al., 2024). Women are also key drivers in addressing the challenges of climate change by cutting greenhouse gas emissions, ending deforestation, increasing the use of renewable energy, and improving access to water and sanitation. By investing in developing social ventures and community-based organizations that pursue sustainable business strategies, African women entrepreneurs improve the social well-being of their communities, stimulate local economic development, and promote innovation in the implementation of environmentally friendly products and services. For instance, women entrepreneurs in agriculture value chain are promoting the planting of climate-smart crops, ecological manageable farming, and healthy vegetables, both producing food and maintaining the earth's health (Kitole & Genda, 2024). Likewise, women entrepreneurs are emerging to provide sustainable energy solutions to rural and other marginalized communities in the renewable energy sector. Besides, the customers of female business leaders also have a positive impact on the environment and involve themselves in social causes such as education and health. Companies that these women own often employ models that seek to employ other women and other disadvantaged groups in a cycle of success that trickles across the society. As women entrepreneurs embark on establishing more sustainable and socially responsible businesses, they act not only as agents of economic growth but also as agents of change for sustainable living standard and environmental stewardship (Abdelwahed et al., 2025).

The purpose of this chapter is to introduce the reader to the complex and ever-changing role of African women in particular as entrepreneurs who are continuously embracing change as they spearhead change across the continent. It showcases their efforts toward influencing environment, society, and economy pointing out how their operations and models are spearheading change toward a better world. Using live cases from different African countries, the discussion will highlight women entrepreneurs in various sectors who are implementing sustainable business practices, which will also be illustrated as creative. The chapter will also discuss success stories of African female entrepreneurs adopting technology and innovation as key tools to address challenges and make a greater impact when it comes to sustainable energy, farming, and education. These women have not only been working tirelessly to ensure the achievement of some of the SDGs that affect the African continent but are also acting as pioneers of change to entrepreneurship in Africa. However, the chapter will also look at the organizational dynamics that hinder women entrepreneurs in Africa and how they are coping with the challenges.

It will explore the elements that relate to supporting ecosystems and financing availabilities and gender-sensitive policies that pave the way to empower more women to participate in the entrepreneurship ladder. The topics of gender, sustainability, and entrepreneurship will be discussed as the intersectionality will demonstrate how African women are integrating social, economic, and sustainability goals in their ventures. Although this chapter emanates from an African context, the insights derived from the African female entrepreneurs apply to conscious entrepreneurial ventures worldwide. Thus, the various systems and strategies employed by the African women entrepreneurs provide insights into sustainable solutions to world problems. Their cases showcase how entrepreneurship is not only an economic opportunity but also a way to transform the environment and society globally.

The entrepreneurial diversification that is evident in Africa, especially for women, has potential impacts on global sustainable development objectives. At a time when the developed and developing nations are facing challenges like climate change, scarcity of natural resources, and disparities prevailing in societies, the strategies of African women entrepreneurs have the potential of inspiring other continents. This research indicates that African women have realized that it is possible to engineer growth that is economic, sustainable, and relevant, not at the cost of the environment but by harnessing local knowledge and coupling it with modern technologies. Women entrepreneurs of Africa should be seen not only as players in building the future of Africa but also they are playing the role of a global citizen and contributing to the enhancement of the quality of life on the planet. Their efforts, determination, and focus on sustainability are testimony to the fact that entrepreneurship can be an effective agent of change in the long term, not just locally, but globally. Hence, this chapter will present their experiences, approaches, and effects, providing a clear view of how the African women entrepreneurs are agents of change toward a sustainable future.

2.2 THE INTERSECTION OF GENDER, DIVERSITY, AND ENTREPRENEURSHIP IN AFRICA

2.2.1 Challenges Faced by Women in Traditionally Male-Dominated Industries

Today, gender diversity is on the rise in Africa with women venturing into leadership and ownership of their business across different industries, but the challenge remains—gender inequality is still rife, especially in traditionally male-dominated industries and sectors. Women entrepreneurs in Africa experience various hurdles while operating in sectors heretofore dominated by men like construction, manufacturing, technology, and energy among others (Wu, 2013). These industries are characterized by the patriarchal model of business, where a variety of managerial and entrepreneurial decisions are made by men, the circulation of access to markets is controlled by men, and business networks are male-only. The women participating in these sectors are rarely taken seriously, and many face prejudices and discrimination within the institutions they work. For example, obtaining funds for women-owned businesses operating in sectors dominated by men is still problematic in Africa. By a report of the World Bank, women entrepreneurs in Africa attract less capital compared to their male counterparts as the sources of funding consider women's businesses as higher risks, irrespective of their prospects or financial histories (Maziriri et al., 2023). This imbalance is even more significant in industries such as technology and engineering, as capital investments are crucial for companies' growth. Women who go out to look for investment end up being marginalized by male-dominated venture capital circles, this further hinders their ability to expand their ventures and come up with innovative solutions within their respective fields.

Furthermore, women who venture into sectors that were previously dominated by men are most likely to be confronted with gender stereotyping that puts into question their ability to effectively manage businesses. These misconceptions, where men are seen to be more capable when it comes to technical or physically challenging work, lead to harassment against women in the workplace. They are evident not only in organizational cultures but also in overall cultural beliefs that dissuade women from seeking jobs in these industries (Hanson, 2009). Hence, they are doubted by clients, suppliers, and even employees, and this demoralizes them and affects their management ability. However, there are also various challenges for women entrepreneurs in

Africa; these are the challenges that many women business owners are overcoming bravely. Some of them are establishing the system of support by women's business association and mentoring programs to stand united on combating gender discrimination. Some are inventions which are radical and revolutionary in nature and redesigning existing markets by demonstrating that women can excel in fields which have traditionally been monopolized by men. This evolving status of women into becoming successful entrepreneurs is gradually shifting the social paradigm and the traditional norms that associate women with leadership positions in sectors that are dominated by men though there is still a long way to close these gender gaps (Vundi, 2020).

Cultural, social, and economic constraints remain stiff hurdles to women aspiring to undertake business in Africa. Among many African cultures, women are expected to remain at home and take care of the family and the home without focusing on their careers. Such cultural practices often constrain the time, mobility, and funds women have to spare to their enterprises. In clinics, schools, and family settings, women are expected to provide care to children and household chores which rarely leave them time to embrace entrepreneurship. This division of labor may limit their opportunities to engage in the economic activities or opportunities and slow down the growth of business run by women. The other social factors that hinder women from accessing the resources they need to start and grow businesses include social norms of ownership and inheritance. In some African societies, the women cannot even inherit property or land which is useful as security for credit or other forms of financing. Denied access to formal credit sources without collateral, women have limited access to financial instruments, which provide less money with high interest rates in the informal market. Further, in patriarchal cultures, women's control over money is often limited, and men and boys in families or marital relations dominate the acquisition and decision-making in business (Kumar, 2024). However, cultural and social practices form only a part of the problem; the economic disparity that exists worldwide amplifies the challenges that women face as entrepreneurs. The poverty situation in Africa also sees women being more vulnerable to the effects than men due to limited access to education and training. Lack of education to quality standards means that many women entrepreneurs are less qualified, less experienced in business and thus have inferior chances of facing tough competition in such sectors. However, due to the difficulty that is associated with accessing capital, high establishment costs in sectors such as retail that are predominant among female entrepreneurs, and the lack of access to business education and coaching, most women are unable to grow their businesses beyond the micro or small business level. Local and international governmental as well as nongovernmental organizations are gradually appreciating the importance of eradicating the aforementioned vices through polices and other strategies. Programs that provide skills development, knowledge on financial management, business mentoring, and legal aid are useful in helping women overcome these cultural, social, and economic settings. The achievements thus far have shown that women have the potential to surmount these challenges and even succeed in business when amended (Maziriri et al., 2019).

Patriarchal systems significantly mobilize the involvement of women in ventures and impact the business choices made by female African entrepreneurs, sectors they venture into, and prospects of developing and managing their businesses. Gender norms restrict women to certain fields that are particularly aligned to women-like food processing, handicrafts, businesses, and stores, and many more. Although such sectors generate significant revenues within a country's economy, the growth and profitability aspects are usually relatively lower than technology or financial services, for instance, infrastructure industries. This consequently explains why female-owned enterprises are still constrained, and most are limited to micro and home-based enterprises, hence exacerbating gender inequalities in enterprise development. The problem of women's entrepreneurship in societies where traditional gender expectations are strictly observed can be the discouragement of women to engage in high risk and or innovative business risks that could involve expansion beyond conventional female roles (Osunmuyiwa & Ahlborg, 2019). The imposition of the role that women should not take risks and concentrate on businesses with a limited turnover can slow development and restrict their contribution to the economy. Such culture can also influence women's attitude toward themselves and their capacity to perform and lead business, many women absorbing society's doubts about their effectiveness in some industries. However, women who are able to challenge these gender roles and establish themselves as business owners are more inclined to practice sustainable, socially responsible, and communal forms of business. For instance, many female entrepreneurs across Africa have seized the opportunity of being caregivers and leaders in their households and societies to factor social and environmental responsibilities that enable them to empower not only themselves economically but also the society at large (Ndou et al., (2024).

2.2.1.1 Diversity as a Driver of Innovation

The current and increasing focus on diversity is important because it is an essential driver in any setting but especially in the entrepreneurial environment. In Africa, for instance, the sociocultural, ethnical, and economical diversity of the societies of most countries means that women entrepreneurs come from diverse backgrounds. And that plurality is a key to innovation and allows women entrepreneurs to come up with solutions to numerous issues they face. It has been found that when gender diversity is encouraged in entrepreneurship, it also provides new ideas for thinking that cannot happen in other more closed groups. Different categories of entrepreneurs found that women business owners tend to solve problems differently than male ones, especially concerning social or environmental challenges. For instance, in the renewable energy industry, female-owned companies in Africa are deploying clean electrification services to rural and peri-urban communities since women and children are most affected by energy scarcity. It is more nuanced, gendered, and therefore perfect for targeting isolated social groups and for promoting broader economic development. Apart from gender, ethnic and socioeconomic diversities are also important factors to innovation in African entrepreneurship. Diverse ethnic knowledge regarding local markets, culture, and consumer behavior of the two entrepreneurs also enables target market niche adaptation by entrepreneurs (Al-Qahtani, 2022). Socioeconomic diversity, especially for women, guarantees that innovation is not conducted with the support of massive capital. Women from poor settings have come up with innovations that are cheap since they use meagre resources to address social and environmental challenges innovation. The integration of diversity into entrepreneurial ventures also extends to better developing more sustainable and flexible businesses or organizations. Hypotheses also argue that multi-teams' groups possess stronger risk perception and more effective risk response mechanisms. Several women entrepreneurs set the wheel of their business decisions as they battle with multiple challenges in their everyday lives and careers and thus are more times wiser and capable of change whenever the odds are stacked against them.

An exploration of the various unique women-led businesses across African countries entails that cultural disparities, policies, and economic factors do affect the entrepreneurial results differently across the regions. For instance, in North Africa, women in business face more challenges than their counterparts because they work in societies that frown upon women engaging in economic activities. However, women-led businesses in this area have found a niche in the textile and food industry as well as in the production of traditional arts and crafts in the creation of high-quality value-added products for domestic and export markets. On the other hand, levels of women engagement in business are even higher in the sub-Saharan region which comprises a more diverse environment. Many female entrepreneurs across the African nations such as Ghana, Nigeria, and Kenya are diversifying and innovating, from e-commerce and software development to farming and financial services. The technological advancement such as the use of digital platform and mobile technologies makes it easier for women to produce goods and services, selling them on the international markets, getting micro finance, and discovering new sustainable business models. On the other hand, in Southern Africa, women-dominated businesses are largely found within the formal business environment and are well supported by governments and nongovernmental organizations.

2.2.1.2 Entrepreneurial Women in South Africa

They operate in renewable energy, health, and education sectors among others. Gender policies of a certain country are more liberal compared to others, there are well-developed infrastructure and access to capital, especially for women involved in entrepreneurial activities where women entrepreneurs have the capacity to grow such business models at a faster pace compared to other regions (Haugh & Talwar, 2016). There are differences in the kinds of sectors and challenges that women entrepreneurs encounter across the African continent, but there is some universality—namely, the importance of diversity. These women entrepreneurs, situated in various regions, are using their different experiences to establish enterprises that can solve social, economic, and environmental issues and create the desired change for a better future of the continent.

2.3 SUSTAINABILITY AND WOMEN ENTREPRENEURS IN AFRICA

Women entrepreneurs in Africa are making significant strides toward achieving the SDGs by focusing on sustainability through their businesses. Their contributions are particularly notable in several key areas.

2.3.1 Key SDGs Targeted by Women Entrepreneurs

Women entrepreneurs have a great impact on the achievement of the SDGs in Africa. Their business initiatives have a positive impact on specific SDGs, such as gender equality, decent work, economic growth, and climate action. In the same way, the efforts more so by African women in business create an environment of economic development hence contribute to meeting global agendas for sustainability. One of the key SDGs that women entrepreneurs in Africa actively address is SDG 5: Gender Equality. Gender disparity still exists throughout African countries, thereby, women-owned businesses go a long way in eliminating gender disparity. Many of these businesses give other women an opportunity to work and be trained and be given an opportunity to have leadership positions in their firms. The African women entrepreneurs also rebel against the traditional norms of the society, encourage other women to begin businesses, thus empowering the women of the society to become the role models to emulate. It is by creating businesses that promote gender equality, women entrepreneurs eradicate systems that have for a long time locked out women from mainstream commerce and society.

SDG 8: Another area which has a massive stroke from women entrepreneurs in promotion of Decent Works and Economic Growth. African female entrepreneurship is important since such enterprises participate in the local and national economy by providing employment opportunities to individuals within various regions of the continent. Most women are involved in small- and medium-sized enterprises (SMEs), which are significant agents of economic development in African countries. SMEs commonly contribute to the overall private sector in most African countries, and the role of women's entrepreneurship is critical for realizing sustainable development of economy. These businesses ensure decent work for women besides assisting them become financially independent, and the companies also assist in the economic upliftment of communities and its society.

SDG 13: Climate Action. Most African women are fully conscious of the environmental vices that are so rampant in their societies, such as climate, change, deforestation, and depletion of the natural resources. Women entrepreneurs in Africa are shifting their interest toward creating products and service solutions that do not harm the environment more than they need to. They have businesses ranging from renewable energy companies and efficient farming techniques that are friendly to the environment among others, and they are the emerging green heroes of the continent. They ensure ecological responsibility within the business sphere as well as foster greater stability in their communities throughout years into the future (Aluko et al., 2024).

2.3.2 Real-Life Examples of Successful Women-Led Businesses in Africa Contributing to Sustainability

Several women-led businesses in Africa exemplify the significant contributions to sustainability and various SDGs. One such example is Solar Sister, founded by Katherine Lucey and expanded through a network of African women entrepreneurs in Nigeria, Uganda, and Tanzania. This organization focuses on distributing affordable solar-powered products, such as lamps and phone chargers, to off-grid communities, directly addressing SDG 7 (Affordable and Clean Energy) and SDG 13 (Climate Action). By empowering local women to sell these products in rural areas with limited electricity access, Solar Sister not only promotes clean energy solutions but also creates income opportunities for women, fostering economic independence. The organization further enhances women's entrepreneurial skills through business training, contributing to local economic growth.

Another exemplary initiative is the Akoma Cooperative Multipurpose Society in Ghana, led by women in the shea butter industry. This cooperative combines traditional knowledge with sustainable practices to produce organic shea butter, contributing to SDG 1 (No Poverty), SDG 5 (Gender Equality), and SDG 12 (Responsible Consumption and Production). The women of Akoma use environmentally friendly methods that preserve biodiversity while ensuring fair trade practices that empower women economically. Their commitment to sustainable harvesting techniques protects the shea trees, which are vital to the ecosystem, thereby demonstrating how women-led businesses can create economic value while safeguarding natural resources.

Finally, Green Energy Biofuels, founded by Eziaha Uko in Nigeria, illustrates how women entrepreneurs can address sustainability challenges by advancing SDG 7 and SDG 13. The company produces biofuels from waste materials, such as used cooking oil and agricultural waste, providing an eco-friendly alternative to fossil fuels and reducing greenhouse gas emissions. This initiative also tackles health issues caused by indoor air pollution, which disproportionately affects women and children in Nigeria, by offering cleaner alternatives to traditional fuels like firewood and charcoal. Green Energy Biofuels showcases how women entrepreneurs are leveraging innovative solutions to combat climate change while improving community well-being.

African women in business are setting pace on the sustainable business models incorporating environmental consciousness and social relevance as captured in the SDGs. MitiMeth is another good example of this approach created by Achenyo Idachaba in Nigeria which uses aquatic weeds that are considered invasive species and turns them into baskets and furniture. This eco-friendly initiative not only addresses the environmental challenge of water hyacinth infestation but also creates job opportunities for local communities, particularly women and youth, thus contributing to SDG 8: Decent Work and Economic Growth and SDG 12: Responsible Consumption and Production. Likewise, Kasha, by Joanna Bichsel in Rwanda and Kenya target women and girls' health by selling reproductive health products and affordable menstrual products that are delivered through an online platform and women retail vendors. By empowering women as community sales agents, Kasha addresses SDG 3: Good Health and Well-being as well as SDG 5: Gender Equality in the delivery of essential health products for underserved populations especially the vulnerable groups in low-income and rural areas.

Further, women entrepreneurs are leading the shift toward circular economy systems which are characterized by decrease in waste, using materials more than once, and restoring natural processes. The agricultural producers in Kenya and Tanzania are using crop residues, animal manure, and other organic waste to produce compost or biogas to improve soil fertility and reduce operating expenses, as well as mitigate the adverse effects of climate change. An example of sustainable practices within fashion business includes Vivienne Taa of Kente Gentlemen in Côte d'Ivoire who employs sustainable fashion whereby she reuses material that otherwise may cause an environmental hazard to fashion ivory tower. Thus, such women contribute to creating circular economies that benefit people and the planet in addition to promoting environmental sustainability. In their earlier studies, the author highlighted various forms of innovations leveraged by women entrepreneurs in Africa who are the key contributors to the achievement of SDGs for the future prosperity of the continent (Rosca et al., 2020).

2.4 WOMEN ENTREPRENEURS IN KEY SECTORS DRIVING SUSTAINABILITY

African women entrepreneurs are pioneering innovative approaches to sustainability across key sectors such as technology, agriculture, health, and education. Their efforts are not only driving economic growth but also advancing environmental preservation and social equity. This section highlights how women in these sectors are leveraging entrepreneurship to create meaningful impact, using real-life case studies to illustrate their contributions.

2.4.1 Tech and Innovation

Tech-savvy African women are using technology in different ways to solve social and environmental issues as the technology industry advances. Women-led firms are using technology to provide value-added services in sourcing and managing resources while enabling sustainable solutions (Niethammer, 2013). For example, Onyeka Akumah co-founded Farmcrowdy, which is a digital agriculture platform that brings together investors and small-scale farmers in Nigeria. This tech-driven business not only improves agricultural practices but also creates economic opportunities for women in rural areas, contributing to SDG 2: Zero Hunger and Gender Equality which is SDG 5.

2.4.2 Agriculture and Agribusiness

Horticulture is still a critical component of the African economies, and women are instrumental in the production of food crops for both domestic and export markets while adopting sustainable farming practices. Sustainable practices like organic farming using green products and agroforestry by women engaging in agribusiness contribute to fighting climate change while making food available to consumers. A noteworthy example includes AgriProtein, which was started by Adriana Marais, focusing on a circular system of disposing organic wastes, where possible, by turning them into protein feed for livestock. This innovative approach supports SDG 12: Responsible Consumption and Production and SDG 2: Zero Hunger, improving the food system.

2.4.3 Health and Education

In the health and education sectors, the women of Africa are playing important roles of improving the livelihood of their community by promoting sustainable business solutions. Temie Giwa-Tubosun's LifeBank is a mobile application that matches donors with the needed blood to hospitals to try and prevent shortage of supplies in Nigeria. This initiative supports SDG 3. Another goal is the promotion of good health and well-being and SDG 9 which embraces industrialization, innovation, and infrastructure. In the field of education, Moringa School, started by Audrey Cheng, is producing revolutionary changes in the facet of the technology education in Kenya through offering comprehensive training programs so that the students can get adequate training on digital skills. By bridging the gender gap in tech education, Moringa School contributes to SDG 4: Quality Education and Sustainability Development and Goal 8: Decent Work and Economic Growth.

Thus, women entrepreneurs in their startups carry out efforts to solve some of the shades of African issues in their initiatives while contributing to the achievement of SDGs globally. They lead in technology, agriculture, health, and education sectors proving that female-focused businesses contribute to the development of sustainable economies when given an opportunity to lead. Through defining industries and transforming communities, these women are key agents of positive change and indispensable in Africa's sustainable development.

2.5 CHALLENGES TO SCALING WOMEN-LED VENTURES IN AFRICA

While African women entrepreneurs are driving significant change and contributing to sustainable development, they face several challenges in scaling their ventures. These obstacles are rooted in structural, financial, and societal factors that disproportionately affect women. Despite their immense potential, women entrepreneurs in Africa often encounter barriers to accessing finance, navigating policy frameworks, and overcoming social and cultural constraints. This section will explore these challenges and examine strategies that can help women entrepreneurs overcome these hurdles.

2.5.1 Access to Finance and Funding—Gender Disparity in Access to Credit, Funding, and Investment

One of the biggest barriers that women face in their endeavors is access to finance, particularly in Africa. Gender inequalities in the financial industry have made it difficult to finance women-owned companies. Women are known to be prejudiced by the financial institutions given the belief that women are higher-risk individuals compared to men. Thus, women entrepreneurs often have problems with obtaining conventional credit in banks, venture capital, or other types of financing. The credit constraint is further magnified by restricted asset ownership especially along the gender divide. There are limited surveys on the women in many African countries. This constraint hampers their progress and development greatly and makes large-scale businesses impossible for them. Thus, females continue depending on own and family's resources or petty informal and credit, which is inapt for business development.

2.5.2 Analysis of Microfinance and Crowdfunding as Tools for Women Entrepreneurs

Due to the aforementioned barriers in accessing formal finance, micro finance institutions and crowdfunding are tools that women entrepreneurs can use to access credit. Micro banking, which offers small credit facilities to persons who cannot access any conventional form of loans, has brought a revolution to women in Africa. Women entrepreneurs and small business owners have been able to start up or expand their business through microloan financing. These loans may also have relatively lower demands on collateral conditions, so it is convenient for women to apply for it. Another means that has also been adopted by women entrepreneurs as a novel source of funding is crowdfunding. Crowdfunding sites like Kickstarter, GoFundMe, M-Changa, Thundafund, and many more enable the entrepreneur to post an idea for a business and get funds from the people. Crowdfunding not only brings a lot of money but also gathers people who are willing to support the idea of an entrepreneur. However, there are still challenges facing microfinance and crowdfunding such as high interest rates on the microloan or the challenge of getting noticed in a stiff crowded market of crowdfunded projects.

2.5.3 Policy and Institutional Barriers—Lack of Supportive Policies and Frameworks for Women Entrepreneurs

African women entrepreneurs launch businesses in hostile environments where policies and institutional structures are unfriendly. Even in many developed and developing countries, there are few or almost no specific polices or programs dedicated to empowering women in business. National policies concerning entrepreneurship, investments, and business development lack consideration for gender, especially for women, especially concerning finance, capacity, and legal frameworks. The current entrepreneurial environment does not support women's entrepreneurship due to the lack of policies targeting the barriers that women face in this field; hence, it continues to support the male-dominated economy. Since the country has not adopted gender-responsive policies, women struggle to register their businesses, access credit facilities, and engage in domestic or export trade. In this regard, the governments should look for ways to make the environment more favorable for women involved in entrepreneurship, for example, offering grants for the women-started companies or organizing the spaces where women entrepreneurs can share their experience and find mentors.

2.5.4 Gender Bias in Legal, Regulatory, and Financial Systems

Another challenge that affects women entrepreneurs in Africa is institutional gender bias. Discrimination in laws and regulations and authorities having to do with ownership of property and inheritance, as well as access to credit facilities, discriminates against women. Some cultures still allow women to undertake business under the supervision of male members of the family, thus restricting their freedom as businesspersons. Also, they deny women equal opportunities for loans and investment opportunities as they fortress the financial institutions against them. The problem is that women are rarely presented in the top management within the financial institutions, therefore, decision-makers may have limited understanding of women's challenges. These biases have to be eliminated systematically and as part of that there is need for gender-neutral laws and policies along with the necessity to promote women to leadership positions in all the organizations.

2.5.5 Social and Cultural Challenges

The cultural and social practices in societies tend to exert more pressure on women engaged in business. Traditional African women are confined to the kitchen and are supposed to be carriers of children, which makes it challenging to work for the growth and expansion of a business. Many female entrepreneurs have both business and family obligations and are always torn between the two responsibilities; therefore, they are likely to spend a limited amount of time and energy in developing their business. The lack of adequate social support products like childcare and paid parental leave contributes to this issue. Thus, women in such conditions have to make a choice between their businesses and family: there is no proper financial assistance. This serves not only the purpose of limiting business development within a certain culture but also sustains gender disparities in business ownership. Nevertheless, African women have not been left behind in developing strategies to counteract cultural resistance and compete within their businesses. Specifically, female entrepreneurial connections, training, and incubators for women entrepreneurs are critical factors because they offer access to capital and business skill development. Also, they provide encouragement and tips on how to overcome cultural and social pressures while building a company. Another tactic is to use technology to develop more open business strategies that are not centrally controlled. Through the operational use of digital tools and platforms, female entrepreneurs can run businesses at a personal level, thereby increasing their chances of accomplishing their goals in the business world while providing care to their families. E-commerce, social media marketing, and mobile money solutions strategies have assisted women to conduct their businesses from their homes; hence, limited time is required to be spent away from their families. It is critical to note that increasing the scale of women-led ventures in Africa faces competition challenges, especially because of constrained financing options as well as existing social and cultural norms. But today with micro finance, crowdfunding, and gender-sensitive policymaking, it has become possible for women entrepreneurs to cross these barriers. Through engagement in efforts to reduce gender imbalances when it comes to financial sector, liberalizing of women institutions and other executive measures, women in Africa are being empowered to economically contribute toward sustainability of their respective countries.

2.6 GLOBAL PARTNERSHIPS AND SUPPORT FOR WOMEN ENTREPRENEURS

Global partnerships and support systems have been critical in fostering the growth and sustainability of women entrepreneurs in Africa (Odeyemi et al., 2024). While these women face numerous challenges, international organizations, public-private partnerships, and global collaborations are helping them overcome barriers and scale their ventures. This section examines how global entities and partnerships have contributed to the success of African women-led businesses, driving sustainable innovation and inclusive growth.

2.6.1 Role of Global Organizations (e.g., the UN, World Bank) in Supporting African Women Entrepreneurs

The UN, the World Bank, and other international organizations have for several decades been implementing program that seek to promote the economic enfranchisement of women (Gupta et al., 2024; Buvinic et al., 2020). These organizations understand that women are key to total developmental aims and objectives, such as economic development, eradication of poverty, and sustainability. As a result, they have come up with numerous programs and funding mechanism meant for women entrepreneurs in Africa. For instance, they have been involved in the UN Women's Economic Empowerment and innovation, where they have been useful in delivering capacity-building programs that enable African women gain the necessary skills, knowledge, and networks that enable women in business. This initiative aims at training trainers, consumers, and producers in topics such as finance, information technology, and management for efficiency and the ability to effectively compete in local and global markets.

Moreover, the UN plays an ongoing role in promoting gender-sensitive policies including legal and institutional frameworks which hound the women entrepreneurs. The World Bank, in particular its Women Entrepreneurs Finance Initiative (We-Fi), has also been very central in mobilizing resources that would support women-led enterprises in Africa. We-Fi also sources funds from governments and development finance institutions, as well as private sector actors to finance, support, and connect women entrepreneurs (Maheshwari et al., 2019). Through this initiative, the problem of limited finance access to women entrepreneurs is solved, only startups with the highest growth prospects are considered, and in addition to the necessary funding, such businesses receive knowledgeable recommendations on how to develop their business further. Such global organizations play a vital role of empowering women through availing both financial and policy support for women's entrepreneurship. Mr. and Ms. Anderson are dedicated to ensuring that women are given equal chance to be trained, financed, and exposed to global market thus empowering them for sustainable development.

2.6.2 Impact of Global Initiatives and Collaborations Focused on Empowering Women in Entrepreneurship

Global initiatives and collaborations, driven by organizations like the UN, the World Bank, and regional bodies such as the African Union, have greatly expanded the reach of support for African women entrepreneurs. One notable example is the SheTrades Initiative by the International Trade Centre (ITC), which aims to connect three million women entrepreneurs to global markets by 2025. This initiative provides women with access to international trade opportunities, helping them scale their businesses beyond local boundaries. By offering mentorship, training, and networking platforms, SheTrades empowers women to break into international markets while building sustainable and competitive enterprises (Beleyi, 2022).

Another notable program is the Vital Voices organization, an international nongovernmental organization that empowers women with funding, including entrepreneurial funding. Vital Voices provides for-Profit partner programs including; business training and mentorship, social entrepreneurship, and sustainable business models (Nelson, 2012). The organization has created an active network of women in leadership positions from across the globe that can mentor and partner with African women in business. Such initiatives assist to narrow the gap between the African women entrepreneur and the global market, thereby assisting her to harness international experience, networks, and capital for sustainability. These partnerships underscore the globalization of economic interconnectedness, where building of women's capacities in one country leads to improvements on the global development agenda for all nations. These initiatives' removal of barriers to global procurement and investment to women entrepreneurs in Africa thus encourages women's economic enfranchisement and sustainable development.

2.6.3 Real-Life Examples of Successful Public-Private Collaborations for Fostering Women-Led Businesses

Public-private partnerships (PPPs) have been instrumental in supporting African women entrepreneurs across various sectors. These partnerships leverage the strengths of government agencies, private companies, and international organizations to provide financial resources and business development support to women-led ventures (Dykes & Jones, 2016).

A key example is the Mastercard Foundation's partnership with the African Development Bank (AfDB) under the Affirmative Finance Action for Women in Africa (AFAWA) initiative. This collaboration aims to close the $42 billion financing gap for African women entrepreneurs by providing access to financial services and capacity-building programs. The initiative has supported the growth of women-led businesses in sectors like agriculture, technology, and manufacturing, fostering both economic growth and gender equality (Anwana, 2024).

Another example is the Digital Skills for Women Entrepreneurs program created by the United Nations Development Programme (UNDP) and Microsoft. This initiative focuses on equipping African women entrepreneurs with digital and technology skills, helping them navigate the increasingly digital global economy. By enhancing the competitiveness of women-led businesses, this partnership has also contributed to the development of tech-driven solutions for sustainability challenges.

The Coca-Cola and UN Women 5by20 Initiative exemplifies how PPPs can foster sustainable business practices. Through this partnership, women entrepreneurs have received support to start and grow businesses in the agriculture and retail sectors, with a focus on sustainability. Many participants have adopted eco-friendly practices, such as sustainable packaging and waste reduction, promoting both economic empowerment and environmental conservation.

In conclusion, global partnerships and public-private collaborations have been instrumental in scaling women-led businesses in Africa, particularly in promoting sustainable business practices that align with local and global development goals. By continuing to invest in these initiatives, the global community can empower African women entrepreneurs to lead the continent toward a more sustainable and inclusive future.

2.7 RECOMMENDATIONS FOR ENHANCING WOMEN'S ROLE IN SUSTAINABLE ENTREPRENEURSHIP

To further strengthen the role of African women in sustainable entrepreneurship, several strategic recommendations can be made to enhance their access to resources, networks, and opportunities for growth. These recommendations are designed to address the specific challenges faced by women entrepreneurs while leveraging the power of global partnerships, public-private collaborations, and local ecosystem support.

2.7.1 Expand Access to Finance for Women-Led Businesses

In this regard, there is literature evidence that African female business owners today grapple with several challenges, but finance was pinpointed as the key hurdle among many. There is a need to enhance the women-centered financial products, including microcredit, venture funding, and impact investment. Self-help groups and the government, especially banks and micro finance institutions, should come up with programs that provide loans at low interest, credit guarantees, and impart on the best ways to manage the credit. Further, digital financial platforms could be further developed to target rural and under-banked women making it easier for them to access savings, loans, and investment products.

2.7.2 Foster Stronger Public-Private Partnerships

Success in encouraging women's entrepreneurship has already been demonstrated by PPPs in Africa. To expand these efforts, governments should become even more engaged with private industries, as well as with international organizations and NGOs. Such partnerships may therefore lead to the development of other more elaborate schemes, including women friendly incubation programs, acceleration and co-working facilities, information and data, and business and management training among others. Further, access to PPPs for the promotion of sustainable business practices and green economy will prepare women entrepreneurs to fund climate solutions and sustainable development.

2.7.3 Promote Gender-Sensitive Policy Frameworks

The governments of the African continent should encourage and ensure compliance with policies that will assist women in business. This can involve the abolishment of legal and institutional requirements that prevent women from owning and managing businesses like cumbersome registration procedures and prejudice. National policies must also address issues of equity in property, resource such as land, education, and trainings. Further, there is a need for governments to regularly interact with women entrepreneurs with an aim of grasping their needs, as these inform policy development.

2.7.4 Strengthen Digital and Technological Capabilities

As the influence of digital business scales up, there is need for female entrepreneurs to acquire skills in digital solutions. The governments and private sector firms can come up with training interventions that enhance the capacities of women to fit and participate in sectors such as digital marketing, e-commerce, cybersecurity, or any form of fintech. Also, the availability of cheap internet and mobile services in the developing and rural regions will enable women in the continent to explore the use of the internet for expanding businesses. It also revealed that financial inclusion through modern solutions like digital financial services and blockchain technologies can facilitate transparent access to capital for women freelancers.

2.7.5 Enhance Global and Regional Networking Opportunities

Indeed, one of the major challenges that the female entrepreneurs encounter is the issue of networks and markets. These global bodies including the UN, the World Bank, and the African Union should also sustain initiatives such as SheTrades and Vital Voices for women aspiring to be traders. National and regional organizations such as the African Union should also encourage cross-border trade and business organizations and forums that support African women entrepreneur through mentorship and sharing of knowledge. They can be used to develop partnerships, share information regarding the best practices to adopt, and enhance market access internationally.

2.7.6 Encourage Sustainable Business Practices

Women entrepreneurs should be encouraged and supported in the use of sustainable practices especially when climate change and sustainability form an integral agenda in the world. It is also suggested that more PPP should be extended to support women in micro and small businesses through developing green business models, as through technical and financial support women engage in sectors such as renewable energy, agriculture, and ecotourism. Governments should also encourage sustainable practices through the reduction of taxes, provision of funds, and subsidy for women sourcing sustainable solutions. Furthermore, international organizations can transfer knowledge and experiences on how circular economy and sustainability standards are to be put into practice.

2.7.7 Leverage Global Partnerships for Long-Term Impact

Global organizations like the UN, the World Bank, and the International Monetary Fund should continue to play a key role in promoting women's entrepreneurship in Africa. They can enhance their current programs by creating more long-term, scalable initiatives that address gender inequality and encourage women's participation in global trade. Partnerships with multinational companies should focus on integrating women entrepreneurs into global supply chains, giving them exposure to international markets and fostering sustainable growth.

By implementing these recommendations, stakeholders across sectors can contribute to a more inclusive and supportive ecosystem for African women entrepreneurs. With enhanced access to finance, digital skills, sustainable practices, and leadership development, women-led businesses can play a transformative role in Africa's sustainable economic future.

2.8 CONCLUSION

Women entrepreneurs are now seen as being more crucial in effecting not only Africa's economic transformation but also its sustainability and development. As demonstrated throughout this chapter, female innovators play an invaluable role in positively influencing and molding Africa's sustainable future through various enterprises that support environmental conservation, social equity, and economic development. Businesses owned by these women in fields ranging from technology, agriculture, education, and health are paving way and driving change for both their society and the global society as they seek to promote sustainability.

The role of African women entrepreneurs cannot be overemphasized, as they contribute significantly to the growth of their respective economies and are known to work under very hard conditions with restricted resources. Despite these challenges, they show resilience and innovation, especially in spheres that entail advancement in green practices like green technology, green energy, and sustainable agriculture. Such projects not only address important environmental challenges but are also instrumental in driving social change for gender equality, eradicating poverty, and creating decent work prospects, all of which form the SDGs.

In this chapter, the authors underline the need to embrace diversity in entrepreneurship for the overall global SDGs. The concept of diversity—from gender and ethnicity to and socio-economic backgrounds—is an important source of discovery. Hearing it from women from different disciplines is always fresh and distinct and their strategies are usually closer to the ground as they are more in touch with society than their male counterparts. Therefore, their businesses are largely oriented toward promoting the inclusive growth and making worthwhile contributions to the realization of the SDGs at both the regional and the global spheres. However, it can be seen that there is still a long way to go to achieve the full emancipation of African women entrepreneurs. The obstacles they encounter—limited financial resources and appropriate networks as well as cultural and institutional conditions—cannot be resolved without the integrated efforts of the public and private sectors. In light of these findings, it is crucial that governments adopt relevant policies that encourage the growth of women entrepreneurs, as well as venture capitalists and investors to consider funding female entrepreneurs.

It will require joint efforts of governments, financial institutions, and development organizations to foster an environment that supports women to generate, grow, and expand their businesses. Among them, it is necessary to mention the action that concerns the increase in the availability of finance. And when it comes to finding the required finance, female entrepreneurs in African countries are burdened with many challenges. Enlarging micro credit programs, venture capital, and other appropriate fundraising techniques such as crowdfunding for women entrepreneurs depending on their need and requirements can help the women entrepreneurs to come up with sustainable business models for revenue generation as well as addressing social and environmental responsibility. In addition, there is a need to establish policies that ensure removal of biases that are usually found in laws, regulations, and institutions to improve the entrepreneurial environment. The government should put in place policies which can help women to undertake business with less restraints for them to grow and expand. According to the discussion made earlier, PPP's should remain an important source of funding, coaching, and market linkages for women entrepreneurs.

As for the future of women's sustainable business leadership in Africa, the prospects are bright. In conclusion, empowered women are taking up the mantle to lead innovative businesses, which augur well for the realization of the economic change in Africa comprehensively. The future can encompass a healthy vibrant community that empowers the female entrepreneurs and values their role as a pillar to sustainable practices. In this future, women shall be at the forefront in finding ways to solving some of the challenges facing the continent, such as climate change, food-deficit issue, economic growth, and social justice among others. Therefore, women entrepreneurs in Africa are subject to being critical stakeholders in the continent's sustainable development. Thus, carrying on with investing in women's potential, equipping them with the appropriate tools, and supporting the relevant policies that would allow women to be on an equal footing with men let the international community make sure that more and more women-led businesses persist opening new opportunities for Africa and the entire world to become better and safer both for women and men.

REFERENCES

Abdelwahed, N. A. A., Bano, S., Al Doghan, M. A., Aljughiman, A. A., Shah, N., & Soomro, B. A. (2025). Empowering women through digital technology: Unraveling the nexus between digital enablers, entrepreneurial orientation and innovations. Equality, Diversity and Inclusion: An International Journal, 44(5), 602–626.

Al-Qahtani, M. S. (2022). Women Entrepreneurship for Sustainability-Social, Economic and Professional Empowerment (Doctoral dissertation, Hamad Bin Khalifa University (Qatar)).

Aluko, O., Adefeso, H., & Esther, F. (2024). Unveiling strength: The saga of resilience and empowerment among women entrepreneurs in Ipetu-Ijesha. Educational Administration: Theory and Practice, 30(4), 3087–3094.

Anwana, E. (2024). African Women-Owned SMEs Post-COVID-19: Toward a Gender-Sensitive Policy Development Strategy. In SMEs and Economic Development in Africa (pp. 33–46). Routledge.

Beleyi, F. (2022). Fostering Prosperity for African Female Entrepreneurs: Opportunities for AU–EU Cooperation in Digital Entrepreneurship Networks. In Africa–Europe Cooperation and Digital Transformation (pp. 228–242). Routledge.

Buvinic, M., O'Donnell, M., & Bourgault, S. (2020). Women's Economic Empowerment in West Africa: Towards a Practical Research Agenda. Center for Global Development Working Paper, 557.

Dykes, B. J., & Jones, C. D. (2016). Public-private partnerships in Africa: Challenges and opportunities for future management research. Africa Journal of Management, 2(3), 381–393.

Falaiye, T., Olubusola, O., Ajayi-Nifise, A. O., Daraojimba, E. R., & Mhlongo, N. Z. (2024). A review of microfinancing's role in entrepreneurial growth in African Nations. International Journal of Science and Research Archive, 11(1), 1376–1387.

Fatimah, A., Ali, M., & Chohan, A. R. (2024). Gender dynamics in real estate entrepreneurship: Empowering femRestate for sustainable development. International Journal of Trends and Innovations in Business & Social Sciences, 2(1), 87–97.

Gupta, S., Wei, M., Tzempelikos, N., & Shin, M. M. (2024). Women empowerment: Challenges and opportunities for sustainable development goals. Qualitative Market Research: An International Journal, 27(4), 608–630.

Hanson, S. (2009). Changing places through women's entrepreneurship. Economic Geography, 85(3), 245–267.

Haugh, H. M., & Talwar, A. (2016). Linking social entrepreneurship and social change: The mediating role of empowerment. Journal of Business Ethics, 133, 643–658.

Kitole, F. A., & Genda, E. L. (2024, May). Empowering her Drive: Unveiling the Resilience and Triumphs of Women Entrepreneurs in Rural Landscapes. In Women's Studies International Forum (Vol. 104, p. 102912). Pergamon.

Kumar, S. (2024). Empowering women entrepreneurs: A comprehensive analysis of their impact in G20 economies. Apex Journal of Business and Management, 2(2), 75–96.

Maheshwari, P., Gokhale, A., Agarwal, N., Makena, A., & Borthakur, S. (2019). The global landscape of gender lens investing. IDRC 2(2).

Maziriri, E. T., Mapuranga, M., Maramura, T. C., & Nzewi, O. I. (2019). Navigating on the key drivers for a transition to a green economy: Evidence from women entrepreneurs in South Africa. Entrepreneurship and Sustainability Issues, 7(2), 1686–1703.

Maziriri, E. T., Nyagadza, B., & Chuchu, T. (2023). Key innovation abilities on capability and the performance of women entrepreneurs: The role of entrepreneurial education and proactive personality. Business Analyst Journal, 44(2), 53–83.

McAdam, M. (2013). Female Entrepreneurship. Routledge.

Nayak, V., & Hegde, K. P. (2023). Paryavarna sakhi: Enabling women-led waste management and social entrepreneurship for sustainable development. AMC Indian Journal of Entrepreneurship, 6(4), 8–20.

Ndou, V., Mele, G., Hysa, E., & Mansi, E. (2024). Empowering Women for Technology Entrepreneurship: Opportunities and Challenges. In International Conference on Gender Research (pp. 250–260). Academic Conferences International Limited.

Nelson, A. (2012). Vital Voices: The Power of Women Leading Change Around the World. John Wiley & Sons.

Niethammer, C. (2013). Women, entrepreneurship and the opportunity to promote development and business. Brookings Blum Roundtable Policy Brief, 37(1–10).

Odeyemi, O., Oyewole, A. T., Adeoye, O. B., Ofodile, O. C., Addy, W. A., Okoye, C. C., & Ololade, Y. J. (2024). Entrepreneurship in Africa: A review of growth and challenges. International Journal of Management & Entrepreneurship Research, 6(3), 608–622.

Osunmuyiwa, O., & Ahlborg, H. (2019). Inclusiveness by design? Reviewing sustainable electricity access and entrepreneurship from a gender perspective. Energy Research & Social Science, 53, 145–158.

Rosca, E., Agarwal, N., & Brem, A. (2020). Women entrepreneurs as agents of change: A comparative analysis of social entrepreneurship processes in emerging markets. Technological Forecasting and Social Change, 157, 120067.

Sánchez-Riofrío, A. M., Faytong-Haro, M., Roa, O., & Valderrama-Álvarez, J. (2024). Enhancing women's entrepreneurship in rural ecuador: A study protocol aligned with the sustainable development goals. Economies, 12(9), 225.

Vundi, N. (2020). Empowering rural women for sustainable development in Kenya, opportunities and challenges: Case of Migwani sub-County, Kitui County, Kenya. Impact: Journal of Transformation, 3(1), 24–35.

Wu, D. (2013). Measuring Change in Women Entrepreneur's Economic Empowerment: A Literature Review. DCED Women's Enterprise Development Working Group.

3 Empowering Entrepreneurship in South African Women with Emotional Intelligence

SLR Approach

Arpana Kumari, Shuchita Singh, Aanchal Amitabh, Amit Kumar, and Ali Haj Khalifa

3.1 INTRODUCTION

One of the key success factors for entrepreneurs is to possess emotional intelligence (EI); this fact is validated by several studies and well established by the extant literature. A high level of EI helps entrepreneurs to establish strong relationship and create positive work culture which enhances team performance.

As per the Global Entrepreneurship Monitor South Africa (GEM SA) report, the supportive environment of South Africa for entrepreneurship is rated much lower than the global average. Despite that there is a rise in the number of early stage entrepreneurs, and this entrepreneurial activity is more prominent among women and young people.

Entrepreneurs play a crucial role in shaping South Africa's economy and future growth by bringing in new ideas, job creation, and reduction of poverty. Entrepreneurs also promote resilience culture through their success stories of grit and determination. At the same time, recent research by the University of the Western Cape states that South Africa has much higher failure rate of entrepreneurial ventures than the global average.

Barbara and Anita (2005) in their study identify the traits of "new generation" entrepreneurs in the African continent. The key traits of these new generation entrepreneurs which include women as well as young men are enhanced social and business relationships, trust among the fellow business members, transparent practices, and a strong commitment to develop intra African commerce. They have a mission to improve business conditions for the private sector in Africa and to promote the regional economic integration by helping to change government policies. Despite

DOI: 10.1201/9781003508953-3

all the limitations, these new entrepreneurs are strengthening the private sector and thereby contributing to the economic growth of South Africa.

Derera et al. (2014) established that in South Africa women entrepreneurs faced a lot of discriminatory practices and did not enjoy the equal opportunities as men. There were a lot of gender-specific barriers, especially in raising capital for business. Apart from this, there were several other barriers and restrictions faced by these women Nyakudya et al. (2018). Among the entrepreneurs the number of women is very less; the main reasons attributed to this are poor education and low skills. Botha et al. (2007) studied the role of education and training in South African women entrepreneurs, who already faced discriminatory barriers. With the introduction of Women Entrepreneurship Programme (WEP), there was an increase in skill transfer and business performance of the women entrepreneurs. For success these women required a support from family and government. Sequeira et al. (2016) studied the status of women entrepreneurs in several developing nations including South Africa. They identified several factors instrumental in contributing to the success of these new-age women entrepreneurs. These factors ranged from economic ease to conducive cultural environment and supportive government initiatives. Ndhlovu and Spring (2009) concluded in their study that the prime factors which are crucial for increasing the role of women in business are education and reducing racism along with the intervention of government and a social transformation. Okeke-Uzodike (2024) stated as the findings of their study that the South African government has offered many policies and initiatives to promote women entrepreneurs.

An innovative mindset and entrepreneurial education always have a positive impact on the success of an entrepreneur (Maziriri et al. 2024). Moreover, female entrepreneurs show more innovative mindset and are highly motivated and continue to pursue their businesses even after facing several challenges. Meyer studied the factors motivating South African female entrepreneurs and concluded that female entrepreneurs are highly motivated to remain in business and pursue growth despite challenges. These insights can guide policies promoting female empowerment, job creation, and economic growth. Female entrepreneurs exhibit considerable amount of EI which guides them through barriers and restrictions, and they tend to form natural groups. Halberstadt and Spiegler (2018) in their paper emphasize the importance of social networks, which may be shaped by the EI of female entrepreneurs in managing these relationships for idea generation and development. Social networks are a significant part of the personal context influencing the idea-fruition process of female social entrepreneurs. Sixaba and Rogerson (2023) in their study "Characteristics, motivations, and business operations of small tourism entrepreneurs" state that women own and operate the majority of small accommodation businesses in the Eastern Cape. The study provides new insights into peripheral tourism entrepreneurs, especially women, and highlights the importance of networking for business success in resource-constrained regions. EI plays a significant role in the success of women entrepreneurs in peripheral regions. Ijatuyi et al. (2022) in the study "Socio-economic Determinants of Female Entrepreneurship" state that educational status, EI, job creation (temporary and permanent), extension services, and responsible use of farming inputs significantly impacted female entrepreneurship development. Ahwireng-Obeng (1993) studied about Women in South Africa and how they face systemic barriers such as limited access to resources, institutional discrimination, and sociocultural constraints. Despite these challenges, there is significant potential for women in small and micro-enterprises. Tailored strategies, including gender-sensitive policies and support systems, can unlock this potential and contribute to socioeconomic reparation. Neneh and Welsh (2022) wrote a paper titled "Family Support and Business Performance" that explores the role of family support in the success of women entrepreneurs. The study found that high levels of emotional and instrumental support from family can help female entrepreneurs achieve high business performance.

This study focuses on the role of EI as one of the key factors in the success of South African women entrepreneurs who not only face discrimination but also lack proper entrepreneurial knowledge and support. The study seeks to explore available literature and suggest major gaps in research to extend the literature on EI with special reference to South African Women's Entrepreneurship.

3.2 REVIEW OF LITERATURE

3.2.1 Emotional Intelligence

Salovey and Mayer defined EI as "the ability to monitor one's own and others' emotions, to discriminate between them, and to use the information to guide one's thinking and actions." Self-awareness, emotional regulation, empathy, and impulsivity are personal skills that Bandura has linked to EI.

The ability to monitor an individual's self-emotional framework and comprehend the emotions of others makes this concept extremely helpful for human resource specialists. It encompasses skills such as emotional awareness, emotional regulation, empathy, and interpersonal relationship management. EI enables leaders to connect emotionally with employees, fostering a supportive work environment that enhances organizational success. It influences decision-making by integrating cognitive and emotional components, leading to more thoughtful and effective outcomes in challenging situations (Sharma et al., 2023).

3.2.2 Emotional Intelligence and Entrepreneurship

Emotionally intelligent leadership fosters healthier work environments and enhances interpersonal skills, which are crucial for entrepreneurship (Jagneet and Bhatia, 2024). Developing EI can enhance the capacity of future leaders in Africa, aligning with national values and promoting participation in the development process. With high EI competency, entrepreneur exhibits self-confidence, trustworthiness, achievement orientation, service orientation, change catalyst, teamwork, and collaboration. Emotional regulations facilitate entrepreneurs to be more innovative (Ngah and Salleh, 2015). The power-packed EI enhances one's stress handling, risk taking, resilience, and decision-making capacity which are quintessential for entrepreneurs. EI and the success of immigrant-owned small businesses in South Africa indicate that EI contributes to both organizational and personal success among entrepreneurs in this context (Olawale, 2019). Neghabi et al. (2011) in their proposed developed intentions model (DIM), which was based on Krueger intentions model, highlighting the concept of opportunity as the core of entrepreneurship, stated that under identical situational variables and intellectual capability, individuals with developed EI would have enhanced degree of self-efficacy even in adverse conditions. While the focus on EI in entrepreneurship is promising, it is essential to consider the challenges faced by entrepreneurs, such as economic instability and access to resources, which can overshadow the benefits of EI in practice.

3.2.3 Emotional Intelligence and South African Women

EI is increasingly recognized as a vital attribute for effective leadership among South African women, particularly in higher education institutions (HEIs). Emotional quotient (EQ) skills, including self-regard and emotional self-awareness, followed by interpersonal EQ skills such as empathy and interpersonal relationships are the major contributing attributes in the case of women in South Africa. In the era of the digital landscapes, there are unique challenges for women entrepreneurs, where EI is essential for navigating emotional and social demands, thus promoting equal participation in e-business (chaidi et al., 2022). The critical role of EI is noteworthy in enhancing leadership effectiveness among South African women, particularly in complex and demanding environments Mayer et al. (2017). The contribution of EI is also emphasized in promoting diversity and inclusion within organizations, particularly for black women, facilitating their advancement in entrepreneurial ventures ("Black Women, Emotional Intelligence, and Organizational Opportunities for Growth", 2022).

3.2.3.1 Methodology

The objective of this study is to undertake a detailed review of the scholarly research related to the domain of EI and entrepreneurship among African women.

The research questions identified for the study are as follows:

RQ1: First, to highlight the status of women entrepreneurs in Africa
RQ2: To explore the factors affecting women's entrepreneurship in Africa

The literature was collected utilizing the resources of Scopus. The analytical exploration encompassed scrutiny of 1,180 Scopus publications. In the specific context of the present study, Scopus facilitated the extraction of roughly 1,180,900 documents based on the defined query. To fetch the data from the sources, the following search equation was used **Emotional Intelligence* and *Women Entrepreneurs* or *Women Entrepreneurship and Africa**. The asterisk (*) was selected to encompass any cluster of characters relevant to the title of the paper. This encompasses all synonymous terms within the subjects of title, abstract, and keywords. The time frame was from 1987 to 2024. The acronym "PRISMA" stands for "Preferred Reporting Items for Systematic Reviews and Meta-Analyses," as exemplified within the network depicted (Figure 3.1). These initially identified records have undergone a refinement process employing the defining characteristics.

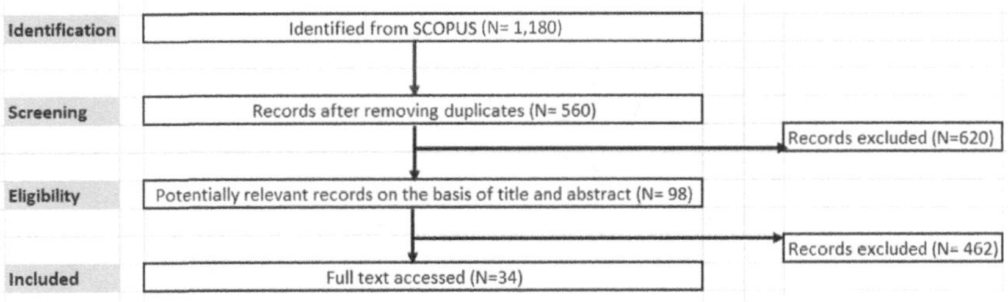

FIGURE 3.1 PRISMA flow chart.

Source: Authors' contribution.

3.3 DISCUSSION AND CONCLUSION

The paper highlights the main research literature available on the role of emotional intelligence with special reference to South African Women Entrepreneurs. Table 3.1 demonstrates that having a high level of EI as a powerful tool can help South African Women Entrepreneurs to face their unique business challenges. The high EI quotient makes them self-aware, and they can better analyze their emotions and take more informed decisions by understanding their strengths and weaknesses. At the same time they can better manage their stress and frustrations as entrepreneurs. With the ability to regulate and manage their emotions they can be better self-managed and exhibit resilience and adaptability to adverse situations. For success of any business, strong relationship network needs to be built where an entrepreneur must exhibit considerable empathy toward emotions of employees, customers, and all the stakeholders. For maintenance of beneficial professional relations an entrepreneur must have social skills such as effective communication and efficient management of conflicts. An entrepreneur must be self-motivated at all points of time to inspire the team members and achieve their desired goals. Thus, with high EI, women entrepreneurs of South Africa can polish their leadership skills, take better decisions, create great work culture, and succeed in their entrepreneurial journey.

The systematic literature review analysis about EI and women's entrepreneurship in Africa has highlighted some major themes like gender disparity, social pressure, family support, funding support, and education opportunities. Each of these identified themes sets up a roadmap for future researchers to dig deeper and enrich the field of the present research, as a theoretical implication of the present study. Also, the study suggests some policy-related points that can be categorized as practical implications of the study.

The paper also identifies future potential for tailormade EI training programs for South African Women Entrepreneurs keeping in mind their unique challenges because such programs can empower these women entrepreneur and contribute significantly to the economic growth South Africa. With government support and favorable policies, suitable financing options, family support, and appropriate training programs, the South African Women Entrepreneurs can be at par with their global counterparts.

TABLE 3.1
Structured Literature Review

S. No.	Theme	Author/s (Year)	Objective	Nation	Data Collection	Sample Size (N)	Analysis/ Technique/ Approach	Findings
1.	New generation African entrepreneurs	Barbara and Anita (2005)	The objective of the paper is to identify the traits of "new generation" entrepreneurs in the African continent.	10 countries of Africa	Interview method	N = 57	-	Findings identify the characteristics of "new generation" entrepreneurs in Africa. The study reveals that they evaluate goals and believe in establishing cross-border networks to achieve organizational goals.
2.	Gendered informal entrepreneurship	Richard Grant (2013)	The objective of this paper is to identify the gender difference in rural entrepreneurship in Soweto.	Soweto	Survey of firms and interview	N = 100 firms and N = 30 firm owners	-	The paper identifies the urban entrepreneurs of Soweto who create zones with rural entrepreneurs and identify them as social capital. But there is a gender bias, and existing policies are not sufficient to address this issue.
3.	Gender and startup capital in Africa	Derera et al. (2014)	The objective of this paper is to address the gender-specific disparity in seeking startup capital in South Africa.	Pietermaritzburg, KwaZulu-Natal, South Africa	Semi-structured interviews and surveys	N = 25	-	The findings disclose that women entrepreneurs in South Africa face difficulty in raising capital for unique businesses. The study also discloses the contribution of women entrepreneurs to the informal economy of South Africa.
4.	South African tourism entrepreneurship	Annie Hikido (2017)	The objective of this paper is to study the disparities between blacks and whites by deeply exploring the denials received by black female entrepreneurs in South Africa.	Capetown, South Africa	Mixed-method approach	-	-	The paper identifies the deep disparities between blacks and whites based on race. This leads to the marginalization of black women entrepreneurs in South Africa.

5.	Women and apparel entrepreneurship	Hodges et al. (2015)	The objective of this paper is to explore the strategies that are being used by female entrepreneurs in Russia, South Africa, and Thailand, operational in the apparel business.	Russia, South Africa, and Thailand	Interview method	N = 11	Thematic analysis	The paper explores the need to be innovative and creative. Female entrepreneurs in the apparel industry need to first identify the demand and trend of the industry and also think about innovative ways of funding to avoid the deficit of capital. Being small-scale business houses, they also need to manage the competition.
6.	Women's entrepreneurship	Botha et al. (2007)	The objective of this paper is to measure the effectiveness of the women's entrepreneurship program (WEM) being run in South Africa.	South Africa	Survey method	N = 180	Factor analysis	The findings disclose the effectiveness of the WEM, as the firms tend to increase their profitability, employability, and productivity after completing training with WEM.
7.	Women entrepreneurs and green economy	Maziriri et al. (2019)	The objective of the paper is to identify how women entrepreneurs are engaged in the green economy.	Gauteng, South Africa	Semi-structured interviews	N = 20	Thematic analysis	The findings of the paper highlight that women entrepreneurs in Gauteng are involved in the green economy by being involved in green education and green production. The study also highlights the need to explore innovative ways to raise green finance.

(Continued)

TABLE 3.1 (*Continued*)
Structured Literature Review

S. No.	Theme	Author/s (Year)	Objective	Nation	Data Collection	Sample Size (N)	Analysis/ Technique/ Approach	Findings
8.	Women's micropreneurship and role of family	Xiong et al. (2018)	The objective of this paper is to understand the role of family and cultural background in deciding the role of women in a family business in extremely poor conditions in the villages of South East Nigeria.	Southeast Nigeria	Focus groups and interviews	N = 2 villages	Thematic analysis	The findings disclose that in the villages of South East Nigeria, women entrepreneurs run their family businesses with limited resources like finances and appropriate skills. The priorities for them are fulfilling basic needs rather than being profitable.
9.	Factors contributing to Women's Entrepreneurship	Sequeira et al. (2016)	The objective of this paper is to study the status of women entrepreneurs in developing countries. The study also aims to identify the major factors contributing toward women entrepreneurs' success in developing countries.	Argentina, Brazil, Hungary, India, Mexico and South Africa	Sample Survey	N = 350 female entrepreneurs	-	The findings disclose that household income, economic ease, a conducive cultural environment, and government initiatives are prime factors impacting the entrepreneurial success of women in developing countries.
10.	Idea fructification and female social entrepreneurs	Halberstadt and Spiegler (2018)	The objective of the paper is to study the reasons behind the lack of ideation among women social entrepreneurs.	South Africa	Mixed-method approach	-	-	The study identifies that the social network structure is crucial for ideation by social entrepreneurs in a developing economy like South Africa. The study adds a new dimension to the research by introducing the role of gender in social entrepreneurship.

11.	Entrepreneurial attributes in manufacturing firms	Darma Mahadea (2010)	The objective of the article is to present the assessment of attributes among male and female micro-entrepreneurs in the urban centers of Transkei.	Transkei, South Africa	Questionnaire-based survey	N = 80 microentrepreneurs	ANOVA and Tukey Test using statistical analysis systems (SAS)	The findings of the study identify that male microentrepreneurs employ more labor than female microentrepreneurs. This may lead to unemployment-related problems in labor-centric countries like Transkei. To address the unemployment-related issues, the entrepreneurs need to adopt a gender-neutral approach.
12.	Entrepreneurial attributes of young entrepreneurs	Fatoki (2014)	The objective of the study is to identify the factors motivating young female entrepreneurs.	South Africa	Survey method	-	-	The study examines how female young entrepreneurs use the basic marketing push and pull strategy. Push strategy is more often used due to the lower intentions of users in the South African region.
13.	Emotional intelligence and women entrepreneurs	Mayer *et al.* (2017)	The objective of the study is to check the emotional intelligence level among the women entrepreneurs involved in higher education institutions.	South Africa	Semi structured interviews	N = 23	Content analysis	The study identifies that women entrepreneurs and leaders in the higher education industry demonstrate a high level of emotional quotient, empathy, problem-solving skills, and resilience.
14.	Fintech and women's entrepreneurship	Abbi and Euphrasie (2022)	The objective of the study is to find out if financial inclusion enables entrepreneurship among female entrepreneurs in the marginalized communities of Africa.	Burkina Faso and Cameroon	Survey method	N = 5,066 and N = 6,826 individuals in Burkina Faso and Cameroon respectively.	Probit model	The findings of the study disclosed that women are on the verge of financial exclusion in both countries, which adversely affects the entrepreneurial strength of the females. Women do not own mobile, and illiteracy rates are also higher than men.

(Continued)

TABLE 3.1 (*Continued*)
Structured Literature Review

S. No.	Theme	Author/s (Year)	Objective	Nation	Data Collection	Sample Size (N)	Analysis/ Technique/ Approach	Findings
15.	Role of family and women entrepreneurs	Neneh and Welsh (2022)	The objective of the study is to identify the role of family to support female entrepreneurs in South Africa.	South Africa	Semi-structured interviews	N = 184	Fuzzy-set qualitative comparative analysis	The findings of the paper highlight that high emotional support from the families of female entrepreneurs is required to ensure better financial performance of their firms.
16.	Women entrepreneurs and medical tourism	Ahwireng and Van (2011)	The objective of this paper is to address the demand-supply gap in the field of medical tourism.	Johannesburg, South Africa	Interview method	N = 320 patients	-	The findings of the study identify that though the patients are interested in medical tourism opportunities in South Africa, the cases of breast cancer are increasing in middle-class women in South Africa due to poor lifestyle.
17.	South African women and business	Ndhlovu and Spring (2009)	The objective of the paper is to identify the factors contributing to the participation of South African women in entrepreneurship.	South Africa	Interview method	N = 335	-	The study concludes that increasing education and reducing racism are two prime factors toward increasing the role of women in business. Government intervention and social transformation are also two major factors in increasing women's entrepreneurship.
18.	Government intervention and women's entrepreneurship	Okeke et al. (2018)	The objective of the study is to examine the government interventions in KwaZulu-Natal for facilitating entrepreneurial activities.	KwaZulu-Natal, South Africa	Systematic literature review	-	-	The findings of the study state that the South African government has offered many policies and initiatives to promote women entrepreneurs in KwaZulu-Natal.

19.	Gender gap in entrepreneurship	Nyakudya et al. (2018)	The objective of the study is to examine the risk aversion nature of individuals for not becoming entrepreneurs.	South Africa	Sample survey	N = 19,469	Quantitative approach, probit model, and a non-linear decomposition technique	The findings of the study explain that the major reason for fewer women entrepreneurs is poor education and low skills among women.
20.	Innovation mindset and women entrepreneurs	Maziriri et al. (2024)	The objective of the paper is to identify the impact of an innovation mindset on women entrepreneurs.	Gauteng Province, South Africa	Structured interview	N = 304	Smart PLS	The findings of the paper reveal that an innovative mindset has a positive impact on entrepreneurial success. Along with this, entrepreneurial education plays a moderating effect between innovation mindset and entrepreneurship.
21.	Factors motivating South African female entrepreneurs	Meyer	To understand and predict factors motivating female entrepreneurs to remain and grow their businesses.	South Africa	Structured questionnaire	510	Descriptive, correlation, and canonical correlation analyses	Female entrepreneurs are highly motivated to remain in business and pursue growth despite challenges. Insights can guide policies promoting female empowerment, job creation, and economic growth.
22.	Institutional and cultural influence on women's micropreneurship	Xiong et al. (2020)	To understand how family and cultural institutions shape family business practices in poor entrepreneurial women's context.	Nigeria (South-East)	Participant observation, focus groups, and interviews	-	Thematic analysis	Family is a powerful institution shaping women's roles and business behaviors. Women prioritize family needs (e.g., children's education) over profit maximization, relying on the family for business and using the business to support the family. Tailored micro-enterprise programs are needed to address cultural norms.

(Continued)

TABLE 3.1 (*Continued*)
Structured Literature Review

S. No.	Theme	Author/s (Year)	Objective	Nation	Data Collection	Sample Size (N)	Analysis/ Technique/ Approach	Findings
23.	Gender differences in business performance in Kenya and South Africa	Lemma et al. (2023)	To investigate the performance gap between female-owned enterprises (FOEs) and male-owned enterprises (MOEs).	Kenya and South Africa	Data from the World Bank Enterprise Survey (WBES)	1,522 firms	Analysis of endowments returns on endowments and gendered performance metrics.	FOEs underperform compared to MOEs due to differences in endowments and their returns. Social feminist theory explains the performance gap. Addressing structural barriers and reshaping social structures can significantly reduce this gendered performance gap.
24.	Transition from non-profit organization to social enterprise	Krige et al. (2019)	To introduce students to the core characteristics of social entrepreneurship, using Santos's (2012) positive theory and analyze resource dependency and financial sustainability.	South Africa	Case study analysis of Sharanjeet Shan and Maths Centre; insights from corporate social investment trends.	Not applicable (case-based)	Application of Santos's (2012) positive theory, Drucker's resource dependency theory, and the funding spectrum.	Social enterprises operate in environments of infinite need, requiring financial sustainability strategies beyond donor reliance. Applying the funding spectrum aids in understanding financial sustainability. Mission drift is a key risk to manage.
25.	Leadership and business model innovation during COVID-19	Scheepers and Bogie (2020)	To understand how Uber South Africa adapted its business model during the COVID-19 crisis while maintaining its core multi-sided platform business and sustainability.	Sub-Saharan Africa (SSA)	Case study analysis of Uber's business model adaptations across seven SSA countries during COVID–19.	Not applicable (case-based)	Evaluation of business model innovations, sustainability approaches, and leadership responses during a crisis.	Uber adapted its platform to support communities, deliver essential goods, and provide financial assistance to drivers during the COVID-19 pandemic. This demonstrates contextual leadership and sustainable business model innovation.

26.	Outsourcing development through BoP entrepreneurship	Dolan (2012)	To explore how BoP initiatives convert poor women into entrepreneurial subjects, reshaping their roles to align with global business through ideological and material practices.	Bangladesh, South Africa, and other regions	Fieldwork with BoP schemes and case studies from various regions.	Not applicable (case-based)	Qualitative analysis of BoP schemes, ideological practices, and global brand engagement.	BoP initiatives aim to produce entrepreneurial traits among poor women, deflecting responsibility for development onto them. While these efforts bring rewards for some, they often serve global brands rather than fostering inclusive capitalism.
27.	Strategies used by successful female entrepreneurs to manage challenges in small apparel businesses.	Hodges et al. (2015)	To explore strategies used by female entrepreneurs to overcome the challenges of running small apparel businesses.	Russia, South Africa, Thailand (3 countries)	Primary: in-depth interviews with female small apparel business owners. Secondary: data on small businesses in the three countries.	11 female small apparel business owners.	Qualitative research design; thematic analysis to identify key emergent themes.	1. Female entrepreneurs had to be highly creative in sourcing capital and resources for starting and scaling their businesses due to limited access to traditional funding sources. 2. Entrepreneurs faced significant competition within the apparel industry, necessitating innovative strategies for differentiation, customer retention, and growth. 3. Women in these markets faced additional barriers such as social and cultural expectations, unequal access to networks, and gender bias. 4. Many of the businesses implemented niche marketing strategies, focusing on custom clothing or tailoring, to stand out in competitive markets.

(Continued)

TABLE 3.1 (*Continued*)
Structured Literature Review

S. No.	Theme	Author/s (Year)	Objective	Nation	Data Collection	Sample Size (N)	Analysis/ Technique/ Approach	Findings
28.	Role of gender, particularly women's entrepreneurship, in addressing socioeconomic disparities and challenges in South Africa.	Ahwireng-Obeng (1993)	To explore gender-specific factors influencing entrepreneurial supply, analyze the impact of socioeconomic reparation strategies and propose frameworks for enhancing women's participation in entrepreneurship in South Africa.	South Africa	existing literature, statistical data, and comparative analyses, including reports	The exact sample size is not explicitly mentioned. However, references to surveys and examples indicate a focus on qualitative insights and statistical data about women entrepreneurs in South Africa.	Qualitative analysis, literature review, and economic data interpretation. Frameworks like Cooper's model of entrepreneurship and sector-specific assessments are used to evaluate gender disparities and opportunities.	Women in South Africa face systemic barriers such as limited access to resources, institutional discrimination, and sociocultural constraints. Despite these challenges, there is significant potential for women in small and micro-enterprises. Tailored strategies, including gender-sensitive policies and support systems, can unlock this potential and contribute to socioeconomic reparation.
29.	Gendered poverty, social entrepreneurship, and sustainable livelihoods	Raniga and Ringson (2021)	Analyze women-led entrepreneurship projects as a means to reduce poverty in female-headed households.	South Africa	Survey with 120 social work students (Sep–Nov 2019) and review of policy reports from 40 students.	Survey: 120; Policy review: 40	Quantitative (spreadsheet); Qualitative (thematic analysis); Sustainable livelihood approach framework.	1. Women entrepreneurs benefit significantly from collaborative networks involving multiple stakeholders, which enhance resource access and support systems. 2. Leveraging existing infrastructure to produce goods was identified as a crucial strategy for women-led entrepreneurship projects. 3. Enhancing technical and entrepreneurial skills is vital for empowering women entrepreneurs and promoting their economic self-reliance.

30.	Viability of informal micro-enterprises in South Africa	Woodward et al. (2011)	To investigate the characteristics that make informal retail businesses economically viable and sustainable.	South Africa	Surveys were conducted in 2007 with 764 informal retail business owners across urban and rural areas in nine provinces.	764 surveys; 449 were analyzed for owner's income, and 476 were analyzed for sales.	Regression analysis	1. Female entrepreneurs face distinct challenges, including limited income growth despite capital access. 2. Male ownership correlates with 27% higher income.
31.	Entrepreneurial opportunities in medical tourism	Ahwireng and Van (2011)	To explore the potential of attracting Africa's middle-class women to South Africa for breast care medical tourism.	South Africa	Interviews with 320 patients and 5 staff members at the Johannesburg Breast Care Centre of Excellence.	325 (320 patients and 5 staff members)	Qualitative interviews were analyzed to assess demand potential and constraints in intra-African medical tourism.	1. Africa's middle-class women represent a growing target market for specialized health services. 2. Effective demand is bolstered by increasing breast cancer incidence due to middle-class lifestyles. 3. Success requires a robust and extensive marketing strategy to tap into this emerging opportunity.
32.		Halberstadt and Spiegler (2018)	To contribute to the lack of research on female social entrepreneurs in South Africa and to explore the influence of social networks and personal context on the idea generation and development processes.	South Africa	Semi-structured interviews	11 female social entrepreneurs	A mixed-method approach combining semi-structured interviews, media analysis, and egocentric network analysis	The paper emphasizes the importance of social networks, which may be shaped by the emotional intelligence of female entrepreneurs in managing these relationships for idea generation and development. Social networks are a significant part of the personal context influencing the idea-fruition process of female social entrepreneurs.

(Continued)

TABLE 3.1 (*Continued*)
Structured Literature Review

S. No.	Theme	Author/s (Year)	Objective	Nation	Data Collection	Sample Size (N)	Analysis/ Technique/ Approach	Findings
33.	Characteristics, motivations, and business operations of small tourism entrepreneurs	Sixaba and Rogerson (2023)	To explore the characteristics and motivations of small tourism entrepreneurs in the Eastern Cape, South Africa, and to understand how they operate in a resource-constrained environment, with a focus on their networking strategies.	South Africa	Survey and semi-structured interviews	79 Black-owned small accommodation businesses and 19 key stakeholders	Qualitative analysis of semi-structured interviews	Women own and operate the majority of small accommodation businesses in the Eastern Cape. The study provides new insights into peripheral tourism entrepreneurs, especially women, and highlights the importance of networking for business success in resource-constrained regions. Emotional intelligence plays a significant role in the success of women entrepreneurs in peripheral regions.
34.	Socioeconomic Determinants of Female Entrepreneurship	Ijatuyi et al. (2022)	To analyze the contribution of the Award Incentives and Competition (AIC) program in enhancing women farmers' economic strength and to identify the determinants of female entrepreneurship development in rural North West Province.	South Africa	Structured questionnaires	156	Quantitative methods, including descriptive statistics and probit regression analysis	Educational status, emotional intelligence, job creation (temporary and permanent), extension services, and responsible use of farming inputs significantly impacted female entrepreneurship development.

Source: Author's creation

REFERENCES

Ahwireng-Obeng, F. (1993). "Gender, entrepreneurship and socioeconomic reparation in South Africa", *The Review of Black Political Economy*, Vol. 22 Issue 2, pp. 151–165.

Barbara, E., & Anita. (2005). "The 'new generation of African entrepreneurs': Networking to change the climate for business and private sector-led development", *Entrepreneurship and Regional Development*, Vol. 17 Issue 1, pp. 17–42.

Botha, M., Nieman, G., & Van Vuuren, J. (2007). "Measuring the effectiveness of the women entrepreneurship programme on potential, start-up and established women entrepreneurs in South Africa", *South African Journal of Economic and Management Sciences*, Vol. 10 Issue 2, pp. 163–183.

Chaidi, I., Papoutsi, C., Drigas, A., & Skianis, C. (2022). "Women: E-entrepreneurship and emotional intelligence", *Technium Social Sciences Journal*, Vol. 30, p. 214.

Derera, E., Chitakunye, P., & O'Neill, C. (2014). "The impact of gender on start-up capital: A case of women entrepreneurs in South Africa", *Journal of Entrepreneurship*, Vol. 23 Issue 1, pp. 95–114.

Halberstadt, J., & Spiegler, A.B. (2018). "Networks and the idea-fruition process of female social entrepreneurs in South Africa", *Social Enterprise Journal*, Vol. 14 Issue 4, pp. 429–449.

Ijatuyi, E.J., Oladele, O.I., Abiolu, O.A., & Omotayo, A.O. (2022). "Socioeconomic determinants of female development in the rural North West Province of South Africa", *Sustainability (Switzerland)*, Vol. 14 Issue 1, pp. 547–562.

Jagneet, K., Raino, B. (2024). 1. "From self-awareness to social skills", Advances in Logistics, Operations, and Management Science Book Series, https://doi.org/10.4018/979-8-3693-7372-9.ch008

Maziriri, E.T., Nyagadza, B., & Maramura, T.C. (2024). "Oh no! all my money is gone": The detrimental consequences of participating in stokvels among women entrepreneurs within the South African township economy", *Journal of Enterprising Communities*, Vol. 18 Issue 5, pp. 1193–1219.

Ndhlovu, T.P., & Spring, A. (2009). "South African women in business and management: Transformation in progress", *Journal of African Business*, Vol. 10 Issue 1, pp. 31–49.

Neghabi, et al. (2011). Emotional Intelligence and Entrepreneurial Behavior: Development of Personal Aspects of Intentions Model. *International Conference on Information and Finance IPEDR* (p. 2). Singapore: IACSIT Press.

Neneh, B.N., & Welsh, D.H.B. (2022). "Family support and business performance of South African female technology entrepreneurs", *International Journal of Entrepreneurial Behaviour and Research*, Vol. 28 Issue 6, pp. 1631–1652.

Ngah, R., & Salleh, Z. (2015). Emotional intelligence and entrepreneurs' innovativeness towards entrepreneurial success: A preliminary study. *American Journal of Economics*, Vol. 5 Issue 2, pp. 285–290.

Nyakudya, F.W., Simba, A., & Herrington, M. (2018). "Entrepreneurship, gender gap, and developing economies: The case of post-apartheid South Africa", *Journal of Small Business and Entrepreneurship*, Vol. 30 Issue 4, pp. 293–324.

Okeke-Uzodike, O.E. (2024). "Entrepreneurship and poverty alleviation: A tale of black African women in Msunduzi Municipality, South Africa", *International Journal of Business and Globalisation*, Vol. 37 Issue 2, pp. 237–257.

Olawale, F. (2019). "Emotional intelligence and success of immigrant-owned small businesses in South Africa", *Academy of Entrepreneurship Journal*, Vol 25 Issue 4, Print ISSN: 1087-9595; Online ISSN: 1528-2686.

Santos, F. M. (2012). A positive theory of social entrepreneurship. *Journal of Business Ethics*, Vol. 111 Issue 3, pp. 335–351.

Sequeira, J.M., Gibbs, S.R., & Juma, N.A. (2016). "Factors contributing to women's venture success in developing countries: An exploratory analysis", *Journal of Developmental Entrepreneurship,* Vol. 21 Issue 1.

Sharma, R., Dhanta, R., & Solomon, D.D. (2023). The Role of Emotional Intelligence in Effective Leadership and Decision-Making in Business Management. In *AI and Emotional Intelligence for Modern Business Management* (pp. 98–112). IGI Global.

Sixaba, Z., & Rogerson, C.M. (2023). "Peripheral tourism entrepreneurs in a resource-constrained environment: Evidence from eastern Cape province, South Africa", *Geojournal of Tourism and Geosites*, Vol. 47 Issue 2, pp. 370–380.

4 Empowering Women through Financial Literacy

A Pathway to Economic Inclusion in South Africa

N.D.N.B. Rathnayake, Amit Kumar Singh, and Vikas Garg

4.1 INTRODUCTION

4.1.1 Overview of Financial Literacy and Economic Inclusion

Financial literacy, the skills necessary to grasp and utilize financial concepts (Lusardi & Mitchell, 2022), is fundamental to a person's ability to make informed decisions regarding finances. Economic inclusion, in contrast, speaks to the level of fair access to financial services, resources, and opportunities needed to partake in economic activity. The increasing global attention to the intersection of financial literacy and economic inclusion is attributed to the need to bridge the economic divide and the acknowledgment of its potential for sustainable development by policymakers, financial institutions, and nongovernmental organizations (OECD, 2023).

Financial literacy is an important factor in economic empowerment, as it can help individuals to manage personal finances, save for the future, invest wisely, and avoid financial pitfalls such as excessive debt or fraud (Xu & Zia, 2022). However, gaps in financial literacy levels by gender, socioeconomic status, and geographic region continue to impede inclusive economic growth. Moreover, the lack of access to financial education perpetuates economic inequalities, especially for women, who may face systemic barriers to financial literacy (Allen et al., 2023).

Financial literacy is increasingly important in developing economies, like South Africa, due to the ambiguities posed by informal labor markets, access to banking services, and the new age of digital finance. In developing economies, including South Africa, financial literacy is particularly vital, given the complexities and limited access to formal banking systems and the expansion of the digital financial landscape. The World Bank (2023) asserts that with a population that is more aware, people would make sounder economic decisions, which further strengthens the economy of the nation and lowers poverty. According to recent

DOI: 10.1201/9781003508953-4

studies, this has been particularly important for low-income and marginalized segments, especially women; tailoring these efforts for these populations has shown significant increases in economic participation and resilience to financial shocks (Kempson et al., 2023).

4.2 IMPORTANCE OF WOMEN'S FINANCIAL EMPOWERMENT

Empowerment of women financially plays a vital role in ensuring gender equality and economic viability. Increased financial literacy among women enables them to make independent economic decisions, invest in businesses, and support their families' well-being, resulting in better social and economic outcomes (UN Women, 2023). While women have seen significant improvement in recent years, globally they still face a disadvantage in terms of financial education, access to credit, and ownership of financial assets (Sahay et al., 2022).

The increase in women in the workforce boosts household security and adds considerably to the GDP of the nation. Women's financial literacy and inclusion have been shown to lead to a $12 trillion (McKinsey Global Institute, 2023) increase in global GDP by 2025. When it comes to economic mobility and poverty, there is a strong potential for improved financial literacy among women to reduce poverty in a country as rife with gender inequality as South Africa, but several issues impede women's financial empowerment. This is such as traditional gender roles, less access to financial services, lower employment rates, and knowledge gaps in formal financial education (Klapper & Lusardi, 2023a). This is the biggest hurdle for women to achieve financial security since financial exclusion only reinforces economic dependency, thereby making it difficult for women to save or create wealth. Research has shown that digital financial tools and mobile banking solutions can be potent facilitators of financial inclusion, especially in rural and marginalized areas (Demirgüç-Kunt et al., 2023a).

Moreover, financial literacy programs designed toward women can transform their economic paths, encouraging financial freedom and lifting families out of a cycle of poverty that spans generations. For instance, women's decision-making autonomy in economic activity as well as their economic resilience are reported in programs that include gender-sensitive financial training (Fernandes et al., 2023). These interventions also contribute to achieving the wider sustainable development goals (SDGs), most notably SDG 5 (Gender Equality) and SDG 8 (Decent Work and Economic Growth), highlighting the importance of inclusive financial empowerment as a pathway to sustainable economic development.

4.3 THE SOUTH AFRICAN ECONOMIC AND GENDER LANDSCAPE

Despite being Africa's second largest economy, South Africa still faces formidable economic challenges like high unemployment rates, income inequality, and financial exclusion (World Economic Forum, 2023). These economic constraints have a disproportionate impact on women, as the structural barriers to women accessing financial and labor markets mean that they make up a high proportion of the informal workforce (Statistics South Africa, 2023).

The gender pay gap continues to be an enormous challenge, with South African women earning around 30% less than men, in similar positions, in the workforce (ILO, 2023b). Women also tend to have less access to credit and financial education, which serves to deepen financial vulnerability. Also consistent with this narrative is the high rate of gender-based financial exclusion as indicated by the persistent lower rate of formal banking among women than men (Demirgüç-Kunt et al., 2023). In rural and township areas, many women depend on informal savings groups (stokvels), which, although beneficial, do not provide the security and growth that formal financial instruments do.

The initiatives to promote financial literacy and inclusion in South Africa have not achieved much success. Cultural and logistical barriers mean that many programs fail to reach marginalized women—the National Financial Education Framework (NAEF) in terms of knowledge development and FinScope South Africa in terms of private sector initiatives (FinMark Trust, 2023)). The COVID-19 pandemic has deepened financial insecurity among women, highlighting the need for targeted financial literacy interventions (World Bank, 2023).

Digital financial inclusion: A means of closing the gender financial gap in South Africa. Mobile banking, as well as fintech solutions, are becoming more accessible to women and enable them to participate in economic activities without dependency on traditional banking institutions (Klapper & Lusardi 2023). Additionally, the establishment of microfinance institutions has opened new avenues for women entrepreneurs to access funding, encouraging entrepreneurial initiatives and independence (Sahay et al., 2022).

However, these developments showcase that improved and sustainable female finance policies and gender-responsive financial education must be strengthened for women across the demographics in South Africa. Partnerships between government, financial institutions, and community organizations can strengthen the reach and effectiveness of financial literacy programs (OECD, 2023). What is equally important is the change in cultural norms which restrict women's financial independence and in its absence a diversity in economic discourse.

4.4 THEORETICAL FRAMEWORK

4.4.1 Key Theories on Financial Literacy and Empowerment

These perspectives inform efforts to promote financial literacy and empowerment, as well as how they can help individuals, families, and communities achieve financial well-being and economic independence. Some of the most dominant theories include the Life-Cycle Hypothesis, Behavioral Finance Theory, and Capability Approach.

The *Life-Cycle Hypothesis (LCH)* is a theory used in both economics and finance that posits that individuals plan their consumption and savings behavior over the life cycle, with the aim of smoothing consumption over their life-course (Modigliani & Brumberg, 2023). Under this theory, being financially literate helps people make better decisions about savings, investments, and even planning for retirement (Lusardi & Mitchell, 2023a). Better financial literacy would enable women in developing economies to better navigate their financial life cycle and not find themselves financially dependent and vulnerable (OECD, 2023).

According to Behavioral Finance Theory, traditional finance theories assume that all investors act rationally, but psychological biases and heuristics affect their financial decisions (Thaler, 2023). Due to social customs, women may have different financial habits than men and often tend to be more averse to risks and have less confidence in investment decisions than their male counterparts (Klapper et al., 2023). This theory underscores the importance of specialized financial literacy programs, accounting for behavioral discrepancies in financial literacy interventions.

Referring to Amartya Sen's Capability Approach, financial literacy is not just knowledge but the actual ability to act on the knowledge (Sen, 2023). This framework aligns particularly well with women's financial empowerment as it addresses the sociocultural enabling factors that need to be in place for women to experience financial autonomy, including access to financial institutions, employment, and policies (Nussbaum, 2023).

4.4.2 Feminist Economic Theories

Theories from feminist economics can be used to analyze the role that gender inequities play in financial literacy and economic participation. Such theories criticize mainstream economic frameworks for neglecting the gendered dimensions of financial access and agency (Elson, 2023). The claim that economic policies and financial systems have always been created in a manner that peripheralize women (Folbre, 2023). The fact that a lot of women perform unpaid work related to care and domestic work means that fewer of them participate in formal financial systems; thus, they have less opportunity to save, invest, or build credit (UN Women, 2023). Such programs that do not consider the structural nature of these inequalities may not empower women through financial literacy.

The Gendered Financial Inclusion Framework (Chant, 2023) further proposes that financial education interventions be tailored toward women's unique financial behaviors and constraints. Such microfinance initiatives aimed at improving women's entrepreneurship and economic independence have shown success in South Africa; however, high interest rates and a lack of financial education continue to pose challenges (Sahay et al., 2023). Intersectional Feminist Theory emphasizes how there are multiple aspects that shape women's financial experience, such as race, class, and geography (Crenshaw, 2023). In South Africa, poor black and rural women are more likely to be financially excluded than urban and white women due to the economic legacy of apartheid (World Bank, 2023). Financial policies need to consider a targeted model with intersectional challenges and include culturally relevant financial literacy programs.

4.4.3 Human Capital and Financial Behavior Theories

Financial literacy and economic empowerment are well based on human capital theories. Becker (2023) developed the Human Capital Theory, which indicates that any investment in formal education, skills, and knowledge (like financial education!) will improve the individual's productivity and economic potential. As a human capital, financial literacy allows individuals to capitalize on their economic potential through financial decision-making (Schultz, 2023).

This is also related to the Financial Socialization Theory, which explains how and what individuals learn about financial behaviors based on social influences like family, education, and cultural norms (Gudmunson & Danes, 2023). Studies show that women tend to have less confidence and engagement in financial planning than men due to socialization differences in financial matters (OECD, 2023). By advocating for early goal-based financial education and mentorship programs, society can begin to address the disparities caused by compounding over time for women followed by its advantages offered in the systems.

Other important perspectives are based on the Theory of Planned Behavior (TPB), which holds that attitudes, subjective norms, and perceived behavioral control affect financial behaviors (Ajzen, 2023). Transactions subject to societal expectations and perceived control over financial resources are implicit in women's financial decision-making (Klapper & Lusardi, 2023). The gender divide in financial decision-making is influenced by prevailing cultural norms, such as traditional gender roles that position men as pivotal financial decision-makers in the family, which may inhibit women's involvement in financial planning.

Financial literacy and empowerment are inherently multidimensional constructs but are contingent on the theoretical perspectives discussed earlier. Where traditional economic theories are primarily prescriptive (explaining how people should behave based on the rational economic model), behavioral finance and feminist economic theories address the role of psychological, social, and structural factors in shaping financial literacy. Finally, to improve upon gender disparity in financial literacy requires a holistic model which not only reaches out to behavioral insights but also gender sensitive policy as well as investment toward human capital. Thus, in the context of South Africa, where financial exclusion is an ongoing issue facing many women, these theories can be used to guide the implementation of effective financial literacy strategies that result in sustained economic inclusion.

4.5 LITERATURE REVIEW

4.5.1 Global Perspectives on Financial Literacy and Women's Empowerment

As the foundation of economic empowerment—especially for women—financial literacy allows individuals to make educated decisions regarding their finances, enhances their participation in the economy, and contributes to increased financial well-being overall (OECD, 2023). Women are found to be less financially literate than men, worldwide, which constrains their ability to partake in wealth-building activities and therefore financial independence (Lusardi & Mitchell, 2023). These differences are pervasive in both developed and developing nations, indicating that gender-specific financial issues must be addressed with targeted interventions in this area (World Bank, 2023).

The Global Findex Database (2023) shows that financial inclusion has improved globally; however, there are still considerable gender gaps. For example, women are less likely than men to own a bank account, access credit, and utilize digital financial services in many low- and middle-income countries (Demirgüç-Kunt et al., 2023). This has often been more pronounced due to varying socio-cultural norms, lower income levels, and limited access to financial education (Klapper & Lusardi, 2023). In addition, the digital divide exacerbates women's exclusion from financial tools online, particularly in rural and marginalized areas (Sahay et al., 2023).

Financial education initiatives have been launched by developed countries like the United States and other nations in Europe, aiming to enhance financial literacy among women, as it is believed to help in reducing economic disparity (OECD, 2023). On the contrary, the presence of well-developed infrastructure, financial institutions, and policy frameworks has contributed to effective implementation of financial literacy initiatives in developed economies, while it was vice versa in developing countries (McKinsey Global Institute, 2023).

4.5.2 Case Studies from Other Developing and Developed Countries, including South Africa

Many countries have successfully implemented financial literacy programs that have had a far-reaching impact in empowering women. One example of this type of program can be seen with the Pradhan Mantri Jan Dhan Yojana (PMJDY) initiative in India, which has contributed significantly to improving women's financial inclusion by promoting ownership of bank accounts and high-quality digital financial services (Reserve Bank of India, 2023). However, challenges continue to exist in terms of consistent financial education and access to credit especially in rural areas.

In *Kenya*, the M-Pesa mobile money system has transformed financial transactions by enabling women to access banking services without relying on traditional banking infrastructure (Jack & Suri, 2023). Research shows that M-Pesa has increased women's access to financial services, collateral for business expansion, household expenditure management, and financial independence (World Bank, 2023).

In *Sweden*, due to the national school policy and political framing on financial literacy and gender, each child, regardless of gender, receives an adequate level of financial education based on the national education curriculum at school (OECD, 2023). This has led to a high level of financial inclusion, diminishing gender gaps in financial choice and decision-making.

In *Brazil*, financial literacy initiatives tailored to women-led businesses have resulted in improved entrepreneurial success and economic stability among impoverished women (IMF, 2023). Nevertheless, even with the positive developments, women's financial empowerment in many regions remains limited by discriminatory lending practices and gender biases (Klapper et al., 2023).

Women in South Africa remain disadvantaged in financial literacy despite numerous initiatives to empower them through financial inclusion. FinMark Trust (2023) highlighted that South African women trail behind men in financial literacy levels— shaping their capabilities to making informed financial decisions. These financial exclusions are due to limited access to formal banking services, cultural constraints, and economic disparities. Moreover, many South African women make use of informal savings systems, like stokvels, which encourage a community-based financial safety net but do not enjoy the security or interest growth or benefits offered by formal banking (UNDP, 2023).

The South African government and financial institutions have implemented numerous monetary educational activities to combat these factors. For example, the FSCA Financial Education Program: Design Features and Strategic Priorities aims to improve financial literacy in the country, with special programs planned for women, especially in rural and low-income settings (Statistics South Africa, 2023). Yet issues of digital illiteracy and restricted financial control continue to hinder women from reaping the full benefits of financial services (OECD, 2023).

Microfinance programs, including those from the Small Enterprise Finance Agency (SEFA), have been beneficial to women entrepreneurs in terms of access to credit, which has allowed for businesses to be developed and economic independence (World Bank, 2023). Nevertheless, high-interest charges and tough lending conditions continue to stifle financial inclusion among women in the country (Sahay et al., 2023). Overcoming these structural challenges will require policy reforms to make financial services accessible and affordable and better meet the needs of women.

4.6 CHALLENGES FACED BY WOMEN IN FINANCIAL DECISION-MAKING IN SOUTH AFRICA

There are several barriers facing South African women when it comes to financial decision-making, such as:

- *Limited Financial Education Research:* This shows that South African women tend to be less financially educated than men, limiting their capacity to make informed decisions about their finances (FinMark Trust, 2023).
- *Access to Financial Services:* Due to lack of traditional infrastructure in rural areas, many women do not have access to traditional banking institutions but use informal financial systems, such as stokvels (UNDP, 2023).
- *Income Inequalities:* Despite some gender equality progress in South Africa, the gender pay gap remains large with women reportedly earning around 30% less than men (ILO 2023).
- *Sociocultural Barriers:* In South Africa, specific traditional gender norms restrict women from making financial decisions at the household level, leading to economic dependency (Klapper & Lusardi, 2023).
- *Digital Exclusion:* The rising prevalence of digital financial services are not reaching many South African women due to a lack of skills and access to resources such as digital banking, mobile money platforms, and investment tools (McKinsey Global Institute, 2023).

4.7 FINANCIAL LITERACY PROGRAMS AND THEIR IMPACT IN SOUTH AFRICA

All these barriers prompted South Africa to initiate various financial literacy initiatives. The key programs include:

- *Financial Sector Conduct Authority (FSCA):* Financial Education Program aims to raise financial literacy, savings behavior, and knowledge of investments, especially among women (FinMark Trust, 2023).
- *Small Enterprise Finance Agency (SEFA):* Microfinance Initiatives offer women entrepreneurs loans and financial education to foster business growth (World Bank, 2023).
- *Digital Financial Literacy Campaigns:* Targeted at helping women acquire the skillset to access digital financial services and mobile banking (Sahay et al., 2023).
- *Banking and Financial Inclusion Policies:* Government-led reforms for stronger accessibility of financial services for underserved populations, particularly women (Statistics South Africa, 2023).

The literature on financial literacy and the empowerment of women emphasizes the crucial role financial education plays in promoting gender equality and driving economic growth worldwide. Furthermore in South Africa, it is an interesting journey to look into these challenges faced by women, as it is a market where systemic economic disparity, traditional gender roles, and limited access to digital financial services come together. Whereas financial literacy programs have shown effectiveness, further dissemination of targeted and gender-sensitive interventions are needed to overcome persistent obstacles. Future efforts should emphasize the importance of improving financial literacy, expanding digital financial services, and implementing policies that promote women's financial autonomy.

4.8 CONTEXT: WOMEN AND FINANCIAL INCLUSION IN SOUTH AFRICA

4.8.1 Economic Status of Women in South Africa

In South Africa, women contribute substantially to the country's economy; however, they are still faced with systemic economic challenges that hinder their financial independence and economic mobility. Statistics South Africa (2023) reports that women make up almost 51% of South Africa's population, but they are disproportionately affected by unemployment, income inequality, and financial exclusion. The gender wage gap is still a critical challenge, having women earning about 30% less than men on average doing the same job (ILO, 2023). In addition, women are overrepresented in low-paying, informal sector jobs, which typically do not have proper access to formal financial services and provide neither job security nor benefits (World Bank, 2023).

This economic plight is compounded for women in rural and marginalized communities. A lot of them rely on subsistence farming, small-scale trading, and informal employment with little-to-no financial resilience (UNDP, 2023). Despite these challenges, women are increasingly entering entrepreneurial ventures and playing an important role in South Africa's SME sector. Yet the limited exposure to financial literacy and capital investment opportunities continues to hinder their success (FinMark Trust, 2023).

4.8.2 Financial Barriers: Education, Access to Credit, and Social Norms

These include the low level of financial education, the inability to gain access to credit and banking services, and the sociocultural norms that prevent women from participating in economic activities.

4.8.2.1 Financial Education

The financial literacy levels of South African Women affect their decision-making informed financial decisions. In South Africa, the levels of financial literacy among women are significantly lower than those of men (OECD, 2023). The basics of finance related to budgeting, savings, investments, and credit management are not well understood by a lot of women. This is especially challenging in rural areas, where learning opportunities may be limited, and finance-specific education is exceedingly rare (Klapper & Lusardi, 2023).

South African Reserve Bank (2023a: p. 6) states that the government has attempted to address the issue through various strategies, such as the National Financial Education Framework (NFEF). However, these initiatives are not sufficient, as statistics show that women's access to such organizations has been diminished (McKinsey Global Institute, 2023). Research shows that by tailoring financial education programs to women's unique needs and combining digital financial tools and community-based learning, we could significantly decrease the knowledge gap.

4.8.2.2 Access to Credit and Banking Services

Women in South Africa have substantial barriers to entry in accessing formal credit and banking services. Almost a third of women in South Africa are either unbanked or seldom use banks (FinScope South Africa, 2023), preferring informal financial arrangements, like stokvels (community-based saving groups) instead. Stokvels give people a sense of financial security but do not offer investment and interest-compounding through services as banks would (UNDP, 2023).

Women encounter discriminatory lending practices, as male-dominated financial institutions are more likely to require collateral or higher credit ratings, which rely on the premise that women do not own any property and do not have a source of secure income (Sahay et al., 2023). Many microfinance initiatives and alternative lending platforms have attempted to tackle this issue, ensuring fair and affordable access to credit remains a challenge (World Bank, 2023).

4.8.2.3 Sociocultural Norms and Gender Roles

These perceptions, deeply embedded in cultural norms, continue to stifle financial decision-making power among women in much of South Africa. In many communities, men are still perceived to be the primary financial decision-makers, which restricts women's control over their finances (ILO, 2023). This mindset is quite traditional, and it is possible that this type of thinking impacts women's ability to invest, save, and achieve financial independence.

Gender-based violence and financial abuse are significant issues, with many women remaining financially dependent on male partners, thus becoming vulnerable to economic exploitations (UN Women, 2023). Financial literacy and inclusion thus represent another important step on the path to breaking the cycle and a critical pillar for sustainable, long-term economic independence for women.

4.8.3 Government and NGO Initiatives for Financial Literacy

The South African government, together with the NGOs and the financial institutions, is taking various initiatives aimed at improving financial inclusion and literacy of women in the country, recognizing that women face economic challenges in the current environment.

4.8.3.1 Government Initiatives

- *National Financial Education Framework (NFEF):* This framework, driven by the South African Reserve Bank, enhances financial literacy among all sectors of society with a particular emphasis on vulnerable groups, such as women (South African Reserve Bank, 2023).
- *Women's Economic Empowerment Strategy (WEES):* Strives to increase credit accessibility, financial literacy, and entrepreneurship available to women (Department of Trade and Industry, South Africa, 2023).
- *Digital Financial Inclusion Strategy:* Promotes mobile banking and fintech solutions that would increase women's access to financial services (FinMark Trust, 2023).

4.8.3.2 NGO and Private Sector Efforts

- *FinMark Trust Financial Inclusions Program:* Economic access and microfinance women in the formal and informal rural sector (FinMark Trust, 2023).
- *Small Enterprise Foundation (SEF):* Empowers women entrepreneurs in South Africa through microloans and business training (World Bank, 2023).
- *UNDP Women's Financial Empowerment Initiative:* Partners with local organizations to deliver financial education, access to capital, and entrepreneurship mentorship to women (UNDP, 2023).

CASE STUDIES OF FINANCIAL INCLUSION EFFORTS

CASE STUDY 1: THE RISE OF STOKVELS IN FINANCIAL INCLUSION

Most South African women resort to stokvels—community-based savings groups—in lieu of formal banking. According to FinScope South Africa (2023), stokvels are currently enjoyed by more than 11 million South Africans, with 60% of them being women. These systems of informal savings offer a degree of financial security and group wealth accumulation. However, there are no effective regulatory protections for members, leaving them vulnerable to fraud and financial mismanagement (UNDP, 2023). By integrating members of the stokvel into the formal banking system, we can offer a safer place for their savings, where it earns better returns compared to traditional savings accounts.

CASE STUDY 2: DIGITAL FINANCIAL SERVICES AND WOMEN'S EMPOWERMENT

Digital financial services are reshaping financial inclusion for women in South Africa. Mobile banking and digital wallets have seen tremendous growth, especially in urban areas. TymeBank, a digital-first financial institution, is eliminating traditional barriers like high costs and complex documentation requirements, enabling women to access their services (McKinsey Global Institute, 2023). Studies show that mobile banking provides better financial management and access to credit for women entrepreneurs (World Bank, 2023).

South African women still face economic hurdles, however, ranging from limited access to credit, financial illiteracy, and deeply embedded gender norms despite great strides in financial inclusion. Though government policies, financial institutions, and NGOs have made efforts to bridge this gap, more work is needed for women to achieve financial freedom. Strengthening financial literacy, growing digital banking, and improving credit access policies are key components to ensure long-term economic inclusion among South African women.

4.9 KEY FINDINGS AND DISCUSSION

4.9.1 Current Financial Literacy Levels among South African Women

Women in South Africa continue to face challenges when it comes to financial literacy, with research showing that a gender gap in financial knowledge and access to financial services persists. On the basic financial literacy index, only 38% of South African women across ages scored compared to the 52% of men. Also, it informs that despite the progress with financial inclusion in recent years, women remain more disadvantaged than men when it comes to accessing formal financial services, as many women have lower levels of education, remain bound by traditional gender roles, and have limited access to credit (World Bank, 2023).

According to OECD (2023), women in South Africa are highlighted as being especially underprivileged in terms of investment literacy, retirement planning, and credit management. Women who live in rural areas and lower-income areas, with over 60% of women in rural areas having a lack of formal financial education (McKinsey Global Institute, 2023), are among the most affected. Evidence of the gender gap in financial literacy can also be observed in the digital finance space, where men use mobile banking and digital financial services more than women (Demirgüç-Kunt et al., 2023).

Despite these challenges, South Africa has made strides in encouraging financial literacy via government programs, nonprofit organizations (NGOs), and digital financial solutions. But there remains a need for more gender-sensitive financial education programs, especially within schools and community-based learning programs (UNDP, 2023).

4.9.2 Impact of Financial Literacy on Women's Economic Participation

To grow as entrepreneurs, employees, and as household financial decision-makers, women need to be financially literate. When women start to understand their finances, they are more eager to invest in businesses and to save for the future, enabling them to make educated decisions about their finances, resulting in a more stable household and financial security (South African Reserve Bank, 2023).

4.9.3 Entrepreneurial Success

High level of financial literacy among women is positively associated with women's entrepreneurship. South African women have significantly more access to the needed financial literacy (Global Entrepreneurship Monitor, 2023), resulting in a 35% increase in the likelihood of them starting their own businesses (if they have undertaken financial literacy training). The SEFA (Small Enterprise Finance Agency) Microfinance Initiative and others like it have been pivotal in providing financial literacy and capital to female entrepreneurs and has shown an increase in business survival rates by 20% (World Bank, 2023).

4.9.4 Household Financial Decision-Making

Financially literate women also tend to better manage household budgets, prepare for emergencies, and save for their children's education, helping to achieve long-term economic stability (FinMark Trust, 2023). Financial literacy training among women has increased household savings rates by 18% and reduced dependence on informal borrowing (ILO, 2023).

4.9.5 Employment and Career Growth

McKinsey Global Institute (2023) study stated financially literate women feel more equipped to negotiate salaries, workplace benefits, and invest in themselves through professional development. Yet gender dynamics in workplace financial education offerings restrict the effectiveness of these benefits and underscore the important role of employers to include financial wellness programs specifically designed for the unique needs of women (UN Women, 2023).

4.10 SUCCESS STORIES AND BEST PRACTICES

The challenges are real, but there are many financial literacy interventions in South Africa which have been very successful in paving the way for financial empowerment and financial behavior change among women.

The following case studies highlight some of the most effective initiatives:

CASE STUDY 1: TYMEBANK WOMEN'S BANKING INITIATIVE

TymeBank is a digital bank in South Africa that has been key in increasing financial access for women, particularly those in rural areas. TymeBank, for example, has given access to over 1.5 million women to open their bank accounts using low-cost digital banking access and financial literacy programs, reducing their reliance on cash-mediated transactions (McKinsey Global Institute, 2023). Women who participate in TymeBank's financial education programme report increased savings rates and mastery of credit.

CASE STUDY 2: FINMARK TRUST FINANCIAL LITERACY TRAINING

The FinMark Trust Financial Literacy Initiative has trained over 100,000 South African women on financial education in their communities and over digital platforms (FinMark Trust, 2023). In addition to theory, the program includes practical financial skills such as budgeting, investment planning, and risk management. In a post-training survey, 85% of participants reported increased confidence in their financial management, and 60% reported an increase in their savings behavior.

CASE STUDY 3: UN WOMEN'S FINANCIAL INCLUSION PROJECT

By integrating financial literacy alongside digital financial solutions, the UNDP's Women's Financial Inclusion Project (2023) is therefore pioneering the integration of access to banking and investment tools for women, ultimately promoting their financial literacy. The initiative has supported thousands of women entrepreneurs, with participants seeing increased revenues and gaining more financial independence.

4.11 GAPS AND AREAS FOR IMPROVEMENT

Though South Africa has seen some successes in advancing the financial literacy of women, there are important gaps requiring attention to support sustainable financial empowerment.

- *Lack of Financial Education in Schools:* Financial literacy forms no part of the national curriculum and, as a result, many young women leave school without the necessary financial knowledge to approach their financial futures (South African Reserve Bank, 2023). This disparity can be closed by integrating financial education in a high school and tertiary education.
- *Lack of Rural Financial Literacy Programs:* Rural women still are bereft of financial education due to logistical issues and lack of resources even after the efforts of NGOs and financial institutions (World Bank, 2023). Extending mobile financial literacy programs and employing radio or television campaigns could potentially extend to rural populations.
- *Women Have Limited Access to Credit and Financial Services:* It is widely known that many financial institutions still have strict credit policies that work against women financing, particularly for those women without capital or regular income streams (ILO, 2023). Financial policies and institutions should focus on gender-sensitive lending and opening microfinance access to women entrepreneurs.
- *Lack of Digital Financial Training:* As digital banking is on the rise, many females do not have the technical skills needed to operate financial applications and digital banking tools (Demirgüç-Kunt et al., 2023). Bringing women and girls into the digital fold through digital literacy programs, in addition to financial education, will ensure women have access to the entire financial system.
- *Financial Wellness Programs for Women in the Workplace:* Most workplaces do not have financial wellness programs designed for women's unique financial challenges (UN Women, 2023), such as financial coaching, investment, and salary negotiation workshops to empower female employees.

In South Africa, financial literacy makes or breaks women's economic inclusion. Despite some progress due to government policies, digital banking provisions, and NGO-run programs, many areas still require attention with regard to financial literacy, credit provision, and digital financial inclusion. It is important to ensure that measures such as strengthening financial literacy programs, the expansion of digital financial education, the integration of financial literacy into schools and workplaces, and reform in lending policy continue to enable economic empowerment for South African women.

4.12 POLICY IMPLICATIONS AND RECOMMENDATIONS

4.12.1 Strategies to Improve Financial Literacy

Improving financial literacy in South African women is integral to empowering them economically and improving their social mobility as well. Research shows that when women are exposed to financial literacy programs that share more relevant and specific information to their economic struggles, there is greater financial independence and economic integration (OECD, 2023). Effective strategies include:

- *Incorporate Financial Education in School Curricula:* Incorporating financial literacy training in primary and secondary education will help ensure young women acquire essential financial skills at a young age (South African Reserve Bank, 2023). In Sweden and Canada, the implementation of financial literacy in the classroom has been successful, resulting in improved habits related to money later on (OECD, 2023).
- *Community-Based Financial Training:* Support training and education institutions and NGOs to deliver locally relevant financial literacy and education programs, especially in rural and underserved communities. One example of the success of a peer-to-peer financial education program is the FinMark Trust Financial Literacy Initiative (FinMark Trust, 2023).
- *Gender-Specific Financial Literacy Campaigns:* Public awareness campaigns should focus on the financial rights of women, investments, and banking options. Customizing financial education for gender-based financial trajectories builds better engagement and sustainable impact (World Bank, 2023).
- *Providing Workplace Financial Literacy:* Employers should ruthlessly provide better financial education programs that include retirement planning, salary negotiation, and investments (McKinsey Global Institute, 2023).
- *Utilizing Digital Platforms for Financial Literacy:* Due to increasing mobile infiltration, financial literacy can be spread efficiently with the help of e-learning platforms, mobile applications, and social media campaigns (Demirgüç-Kunt et al., 2023).

4.13 ROLE OF GOVERNMENT, FINANCIAL INSTITUTIONS, AND CIVIL SOCIETY

Such inclusive financial ecosystems for all, especially for women, require a collaborative effort by government, financial institutions, and civil society.

4.13.1 Government Initiatives

- *Strengthening Policy Frameworks:* Governments need to enforce financial inclusion policies so that women's access to banking, credit, and investment opportunities are prioritized (South African Reserve Bank, 2023).

- *Promote Financial Literacy Programs to Women:* One way to narrow knowledge gap is to introduce state-funded financial literacy initiatives for women, especially in rural areas (FinMark Trust, 2023).
- *Women-Friendly Financial Inclusion:* A policy that ensures financial institutions provides women-friendly finance (OECD, 2023).

4.13.2 Financial Institutions

- *Innovating Female-Centric Financial Products:* Financial institutions should innovate financial products that best serve female customers, including loans with low-interest rates for women uniquely positioned to run small businesses and simplified savings accounts (McKinsey Global Institute, 2023).
- *Expanding Women-Centered Microfinance Programs:* Institutions like the Small Enterprise Finance (SEFA) should scale down low-barrier microfinance options for women, especially small business owners (World Bank, 2023).
- *Promoting Financial Institutions to Conduct Outreach Programs:* Collaboration with NGOs to teach women financial literacy creates awareness and uses formal financial institutions and systems (ILO, 2023).

4.13.3 Civil Society and NGOs

- *Advocacy for Financial Rights for Women:* Civil society organizations must combat gender-discriminatory financial policies and promote policies that guarantee women equality regarding access to banking and financial services (UN Women, 2023).
- *Grassroots Financial Literacy Programs:* NGOs can train on localized financial education programs that are targeted specifically toward marginalized women who may not have access to conventional financial education (UNDP, 2023).
- *Encouragement to Female Entrepreneurs:* Organizations should provide mentorship programs, networking opportunities and venture capital access to promote women-led companies (World Bank, 2023).

4.14 DIGITAL AND TECHNOLOGICAL INTERVENTIONS

Digital interventions are crucial for encouraging financial literacy and inclusion of women. Mobile banking and fintech innovations have filled gaps in South Africa, especially among those in the rural and informal sectors (Demirgüç-Kunt et al., 2023). Some of the key digital interventions include:

- *Mobile-Based Financial Literacy Programs:* Digital banking platforms, like TymeBank and M-Pesa, have been successfully providing women with financial education and banking services (McKinsey Global Institute, 2023).

- *AI-Based Financial Aid:* AI-driven financial education tools, such as chatbots and financial coaching apps, can personalize financial literacy programs based on women's behaviors with money (World Bank, 2023).
- *Blockchain for Secure Financial Transactions:* Implementation of savings and lending platforms based on blockchain can reduce fraud and improve financial security for women entrepreneurs (OECD, 2023).
- *E-Wallets and Digital Payment Systems:* Promoting mobile money services, ensuring women can save and transact securely without depending on conventional banking infrastructure (FinMark Trust, 2023).
- *Fintech Firms/NGOs Partnerships:* Collaboration between financial technology companies and civil society organizations can enhance women's digital financial literacy and scale up access to digital financial services (South African Reserve Bank, 2023).

4.15 RECOMMENDATIONS FOR POLICYMAKERS

To improve the financial literacy and financial inclusion of South African women, policymakers can consider the following recommendations:

- *Require Financial Literacy in Schools and Vocational Training:* One way to ensure long-term financial awareness is to make financial education a mandatory subject in high schools, universities, and vocational training institutions (OECD, 2023).
- *Expand Gender-Inclusive Financial Regulations:* Enforce gender-sensitive financial measures that mandate banks to dedicate a percentage of credit volume to women entrepreneurs (World Bank, 2023).
- *Encourage Digital Financial Inclusion:* Launch national digital financial literacy campaigns that educate women on how to use mobile banking, online investment platforms, and e-wallets (Demirgüç-Kunt et al., 2023).
- *Enhance Access to Microfinance and Low-Interest Financing:* Extend government-backed loan schemes and provide incentives to banks to offer low-interest credit to women-led businesses (McKinsey Global Institute, 2023).
- *Implement Workplace Financial Education Programs:* The responsibility to correct for the much larger knowledge gap in financial literacy must also be shouldered by the corporate sector, from large corporations, through mid-caps, to SMEs, by making financial literacy training for all employees mandatory, with a special focus on women's financial empowerment (ILO, 2023).
- *Strengthen Public-Private Partnerships:* Facilitate and incentivize the collaboration of government agencies, financial institutions, fintech companies, and NGOs to scale financial literacy programs and to ensure their effectiveness (UNDP, 2023).

- *Increase Legal Safeguards for Women in Finance:* Implement laws to protect women from financial discrimination, particularly regarding credit, property, and investment access (UN Women, 2023).

Comprehensive policy, digital financial tools, and collaboration between government, financial institutions, and civil society are essential to improving women's wealth in South Africa. South Africa has an opportunity to close the financial literacy gap and empower more women to play a meaningful role in the economy by spreading education programs, accessing technology, and implementing gender-sensitive policies.

4.16 CONCLUSION

4.16.1 Summary of Key Insights

In South Africa, women's financial literacy is critical to achieving economic inclusion and sustainable development. This research has identified the continued disparity in financial literacy across genders, demonstrating that poor financial knowledge, a lack of access to credit, and sociocultural barriers prevent women from achieving financial independence.

Disparities still exist between financial literacy levels among women and men in South Africa, with the most disadvantaged groups being rural and low-income women (FinMark Trust, 2023). Even though digital banking and mobile financial services are rising, women are still less involved in formal financial systems than men due to educational disparities, cultural financial constraints, and exclusive practices (World Bank, 2023).

To address these concerns, some financial literacy interventions, such as FinMark Trust FinScope, TymeBank's financial literacy project, and the UNDP Women's Financial Inclusion Project, have shown various levels of success with white women but have not focused, and thus cannot speak to, on the success of black women with these tools. Interventions have increased savings rates, improved management of credit, and encouraged women to become entrepreneurs (McKinsey Global Institute, 2023).

To achieve this, policy recommendations include integrating financial literacy into school curricula, expanding digital financial education, reforming credit access policies, and strengthening public-private partnerships for sustainable financial empowerment. Governments, financial institutions and civil society continue to play a key role in creating an inclusive financial ecosystem for women (OECD, 2023).

4.16.2 Final Thoughts on the Future of Financial Literacy and Inclusion

Women's financial literacy and inclusion in South Africa will be driven in the future by innovative technology solutions, proactive policy reforms, and targeted educational programs. There are several emerging trends and considerations that will shape this progress:

- *Digital Financial Inclusion:* Mobile banking, fintech solutions, and blockchain technology all allow women to leapfrog traditional financial barriers and access secure and efficient banking services (Demirgüç-Kunt et al., 2023). However, digital literacy gap is still an important issue to address for women to be able to fully take advantage of these services.
- *Gender-Responsive Financial Policies:* By reviewing and updating existing banking regulations to be more gender-responsive, policymakers can standardize access to credit and safeguard women against financial discrimination while promoting female entrepreneurship South African Reserve Bank (2023).
- *Financial Literacy as a Lifelong Process:* Financial education should not be done through single training but rather as a continuous learning process incorporating schools, workplaces, and community programs (UN Women, 2023).
- *Strengthening the Entrepreneurial Ecosystem:* Women-owned enterprises require financing coaching, venture capital access, and professional development resources to properly grow their organizations (World Bank, 2023).
- *Tackling Societal Barriers:* Preparations for awareness campaigns and legislative frameworks are essential in breaking and reshaping traditional gender roles that limit women toward financial independence (ILO, 2023).

4.16.3 Areas for Further Research

While we have made tremendous strides, much research still needs to be done regarding some key areas of women's financial literacy and economic inclusion:

- *Prolonged Effects of Financial Literacy Programs:* Future studies should assess the enduring effects of financial literacy programs on women's income progression, saving habits, and business success.
- *Digital Financial Inclusion in South Africa:* A construct of cultural taboos: women organizations and self-help groups should hit on gender-specific barriers in knowledge, access, and adoption of digital financial services.
- *More Energy-Efficient Solutions:* Deepen a momentum of work into more adaptive, responsive, and perhaps "gender-sensitive" banking solutions.
- *Household Financial Decision-Making:* Research needs to consider how financial literacy plays a role in women's participation in household financial decision-making and if improved financial literacy provides more autonomy.
- *Intersectionality in Financial Literacy:* Research needs to examine the intersections of race, age, education, and geographic location to deepen understanding on women's financial inclusion and inform strategies for localized and tailored financial education efforts.

Empirical studies have shown that financial literacy contributes towards poverty alleviation, economic growth, and equality of gender, all of which fall under societal concern, hence making women's financial empowerment a social issue. It brings to fore the crucial role that collaboration between government institutions, financial service providers, civil society stakeholders, and fintech will play in bridging the financial literacy gender divide, paving way for an inclusive, financially savvier populace.

Despite some advances, systemic barriers still limit women's financial participation in South Africa. With targeted interventions ranging from digital financial literacy to policy reform and the provision of accessible financial services and tools, the economy can witness the unlocking of the full economic potential of its women. The potential for financial literacy to have a meaningful economic impact on women's lives will require further research and the development of a data-driven policymaking agenda to ensure that financial literacy leads to real-world changes for women and a more inclusive economy.

REFERENCES

Ajzen, I. (2023). *The theory of planned behavior: Understanding financial decision-making*. Journal of Behavioral Finance, 24(2), 89–102.

Allen, S., Carlin, B., & Robinson, D. T. (2023). *Gender differences in financial literacy and its implications*. Journal of Economic Perspectives, 37(2), 89–110.

Becker, G. S. (2023). *Human capital: A theoretical and empirical analysis, with special reference to education*. University of Chicago Press.

Chant, S. (2023). *Gender, urbanization, and financial inclusion: A feminist perspective*. Development and Change, 54(1), 112–134.

Crenshaw, K. (2023). *Intersectionality and financial inclusion: Examining race, gender, and economic access*. Feminist Economics, 29(3), 56–78.

Demirgüç-Kunt, A., Klapper, L., & Singer, D. (2023a). *The global findex database 2023: Financial inclusion and tender disparities*. The World Bank.

Demirgüç-Kunt, A., Klapper, L., Singer, D., & Ansar, S. (2023b). *The global findex database 2023: Measuring financial inclusion and the fintech revolution*. The World Bank.

Department of Trade and Industry, South Africa. (2023). *Women's economic empowerment strategy: Policy and implementation plan*.

Elson, D. (2023). *Feminist economics: Transforming economic policies for gender equity*. Routledge.

Fernandes, D., Lynch, J. G., & Netemeyer, R. G. (2023). *The impact of financial literacy on financial behaviors*. Journal of Consumer Research, 50(1), 76–99.

FinMark Trust. (2023). *Financial Inclusion in South Africa: FinScope Survey Results*. FinMark Trust. https://finmark.org.za/Publications/FinScope_SA_Consumer_2023.pdf

Folbre, N. (2023). *Valuing care work and gender equality in economic development*. Cambridge University Press.

Global Entrepreneurship Monitor. (2023). *Women entrepreneurs and financial literacy in South Africa*. https://www.gemconsortium.org/report/202324-womens-entrepreneurship-report-reshaping-economies-and-communities-2

Gudmunson, C. G., & Danes, S. M. (2023). *Financial socialization: Theory and implications for financial literacy education*. Journal of Family and Economic Issues, 44(2), 98–120.

IMF. (2023). *Microfinance and women's financial empowerment in South Africa*.

International Labour Organization ILO. (2023a). *Gender equality and economic empowerment in South Africa*.

International Labour Organization (ILO). (2023b). *Microfinance and women's financial empowerment in South Africa*. https://share.google/t1ITgzJhOFU6rFkcF

Kempson, E., Collard, S., & Moore, N. (2023). *Financial literacy: A review of best practices and challenges*. OECD.

Klapper, L., & Lusardi, A. (2023a). *Bridging the gender gap in financial literacy: South African perspectives*. World Bank Research Observer, 38(1), 45–69.

Klapper, L., & Lusardi, A. (2023b). *Financial literacy and financial resilience: Evidence from emerging economies*. World Bank Economic Review, 37(4), 231–250.

Klapper, L., & Lusardi, A. (2023c). *Women and financial literacy: Bridging the gender gap*. World Bank Research Observer, 38(1), 45–69.

Lusardi, A., & Mitchell, O. S. (2022). *The economic importance of financial literacy: Theory and evidence*. Journal of Economic Literature, 60(3), 547–587.

Lusardi, A., & Mitchell, O. S. (2023a). *Financial literacy and economic outcomes: Global and South African perspectives*. Journal of Economic Literature, 61(3), 345–369.

Lusardi, A., & Mitchell, O. S. (2023b). *The economic importance of financial literacy: Theory and evidence*. Journal of Economic Literature, 61(3), 345–369.

McKinsey Global Institute. (2023). *Financial inclusion and digital banking for women in South Africa*.

Modigliani, F., & Brumberg, R. H. (2023). *The life-cycle hypothesis of saving: An updated perspective*. Journal of Economic Theory, 48(2), 210–234.

Nussbaum, M. (2023). *Women and human development: The capability approach*. Harvard University Press.

OECD. (2023). *Financial education and inclusion: A global overview*. https://www.oecd.org

Sahay, R., Čihák, M., & N'Diaye, P. M. (2022). *Women and financial inclusion: Economic benefits and policy considerations*. IMF Working Papers.

Sahay, R., Čihák, M., & N'Diaye, P. M. (2023). *Gender and financial inclusion: Addressing the barriers*. IMF Working Papers.

Schultz, T. W. (2023). *Investment in human capital and financial literacy: A modern perspective*. Economic Development and Cultural Change, 72(4), 301–324.

Sen, A. (2023). *Development as freedom: The role of financial capabilities*. Oxford University Press.

South African Reserve Bank. (2023a). *National financial education framework and policy recommendations*. South African Reserve Bank. https://www.resbank.co.za

South African Reserve Bank. (2023b). *Strengthening women's financial empowerment through policy*. South African Reserve Bank. https://www.resbank.co.za

Thaler, R. H. (2023). *Behavioral finance: Understanding financial decision-making biases*. Princeton University Press.

UN Women. (2023). *Progress on the sustainable development goals: The gender snapshot 2023*. Retrieved from https://www.unwomen.org/en/digital-library/publications/2023/09/progress-on-the-sustainable-development-goals-the-gender-snapshot-2023

World Bank. (2023). *Global economic prospects: A critical assessment of emerging markets*. Retrieved from https://www.worldbank.org/en/publication/global-economic-prospects-2023

Xu, L., & Zia, B. (2022). *Financial literacy around the world: An overview*. World Bank Research Observer, 37(1), 15–39.

5 Social Sustainability through Women's Empowerment

A Comparative Analysis of Selected Districts in Odisha

Sitansu Ranjan Swain, Vijaya Rudraraju, Sasmita Nayak, and Abdulsattar Abdullah Hamad

5.1 INTRODUCTION

Strategies for addressing rural development should offer focused, varied approaches while taking into account the many options. The idea of sustainability, or sustainable development, is beneficial for corporate planning of all kinds, both locally and internationally, as well as for policymaking and rural isolation. The English term "sustainable" comes from the Latin *sustinere*, which means "to hold up" or "to endure." Sustainability is defined by *Webster's New Collegiate Dictionary* as "to give support" and "to keep up." The definition of "sustainable" according to the *Chambers 21st Century Dictionary* is "able to be sustained/maintained at a particular level without causing damage." It comes from the fields of economics and environmental science. Strategies for addressing rural development should offer targeted, diverse methods, just as they do when weighing the available options. Sustainability, then, is the use of resources without putting the ecosystem in jeopardy (History, 2014).

It has been demonstrated that the biggest barrier to women working for themselves is a lack of financial resources. Numerous development techniques and tools, including microfinance, education, training, and entrepreneurial initiatives, have been employed to close this gap. A nonprofit that provides small-scale, no-collateral loans to low-income women to assist them in starting businesses and developing individually are very less (Postelnicu & Hermes, 2018). Microfinance gives women more access to education, healthcare, and capital, all of which help them become more independent. Additionally, it lowers poverty, promotes entrepreneurship, and increases social capital.

Women with microfinance have more access to capital, healthcare, and education, all of which contribute to their increased independence. Also, it raises social capital, encourages entrepreneurship, and lessens poverty. It enables people to regain control over their lives and alter their financial situation (Galiè et al., 2019). It is crucial to

DOI: 10.1201/9781003508953-5

remember that a variety of demographic factors may have an impact on microfinance's capacity to advance women's empowerment, disregarding Rahman et al.'s (2009) research.

We examined female empowerment from economic, social, political, and psychological angles using Kabeer's (2005) frameworks for resources, agency, and achievement. The authors also examined the impact of microfinance on overall empowerment. Reduced social and economic inequality for women is the aim of microfinance organizations. They offer a range of services, including affordable loans, skill-focused training, chances for savings, and easily accessible healthcare. By providing financial services including microcredit, insurance, training programs, and savings accounts, microfinance seeks to end poverty and give people more power—especially to women. Microfinance is vital in lowering poverty and inequality because it has grown into a vital instrument for providing underprivileged groups in developing countries with a variety of financial and nonfinancial services.

The creative banking and finance approach primarily benefits the poor people in the country. It increases their chances of becoming self-employed for farmers, business owners, and the poor unemployed who struggle to get traditional credit due to their low income levels and lack of collateral (Raphael & Merma, 2017). Microfinance is a successful technique for alleviating poverty, according to governments and development organizations (Agarwal, 2016). Their restricted ability to make money further restricts their financial options. Furthermore, many women encounter difficulties in locating suitable and beneficial financial services. Microfinance has gained popularity as a strategy for ending poverty and empowering the disadvantaged, particularly women (Basu, 2006).

Khurda is ranked twenty-fifth in terms of area and fifth in terms of population. There are 18 lakh residents, of which 52.57% are men and 47.33% are women. There is a notable gender disparity in literacy, with the male rate being 87.90% and the female rate being 70.36%. The overall literacy rate is 79.59%. The district of Khurda spans 2,813 square kilometers. Sambalpur is ranked ninth in terms of district area and twenty-first in terms of population. There are 1,168,217 residents—49.3% women and 50.7% men. There is a notable gender disparity in literacy, with the male rate being 84.35% and the female rate being 67.93%. The overall literacy rate is 76.22%. The district of Sambalpur spans a geographical area of 6,624 square kilometers.

5.2 PROBLEM STATEMENT

In addition to providing poor women with financial and nonfinancial services, microfinance has shown potential as a tool for empowering women and advancing socioeconomic development. But little is known about the role microfinance plays in empowering women, especially in the Odisha districts of Khurda and Sambalpur. By examining the advantages and effects of microfinance on women's empowerment in the districts of Khurda and Sambalpur, this study seeks to close a knowledge vacuum. Programs and policies aimed at promoting gender equality, economic independence, and social mobility must take into consideration the ways that microfinance benefits women. The results supplement the corpus of information already available on microfinance and women's empowerment by providing numbers that were previously ignored for a given location.

Microfinance institutions (MFIs) may find these insights useful in their efforts to enhance their products and services. That's what Alshebami (2015) asserted. This study not only adds to the body of literature already in existence, but it also clarifies the experiences and shows how microfinance can improve women's socioeconomic empowerment in the districts of Khurda and Sambalpur. It will essentially affect politicians, development experts, and local microfinance organizations. The results of this study will help future efforts that use microfinance to empower women and advance fair district development.

5.3 RESEARCH OBJECTIVE

The purpose of this research is to give a thorough understanding of how microfinance and women's socioeconomic empowerment relate to each other in the districts of Khurda and Sambalpur. It will influence local legislators, development specialists, and MFIs both intellectually and practically. Moreover, it enables to see how women's empowerment raises their involvement in social, political, and home decision-making. Increased participation leads to more inclusive policies that support social justice and sustainable governance by taking into account the objectives and interests of various groups.

5.4 LITERATURE REVIEW

Iezza (2010), Paul (2010), Quayes (2012), Sekabira (2013)

Furthermore, considering the current state of affairs, sustainability is seen as "a way of acknowledging how humankind has created an imbalance." Sustainability calls into question a

> moribund economic system that has drained the world of many of its finite resources . . . generated a meltdown in global financial systems, exacerbated social inequality in many parts of the world, and driven human civilization to the brink of catastrophe by unwisely advocating for economic growth at the expense of resources and essential ecosystem services.

This is in addition to sustainability's fundamental effects on ecology and the natural environment. There aren't many differences between rural development and sustainable rural development. The technological, environmental, and economic challenges that face rural communities are also highlighted. Therefore, the rural population's daily necessities.

Kar (2012), Kimando et al. (2012), Kipesha and Xianzhi (2013), Janda and Turbat (2013)

Understanding the connection between MFI financing and financial sustainability is an understudied field in the field of microfinance. The factors influencing financial sustainability and its results, as well as the relationship between capital structure and MFI performance, have been the main areas of attention for this study.

Brau and Woller (2004), Borbora and Sarma (2007), Paul (2010)

Following an analysis of prosperous rural credit initiatives in developing nations, Ohio State University arrived at two important findings: (1) Institutional sustainability required financial self-sufficiency; and (2) institutional sustainability was necessary for the effective provision of financial services to the underprivileged. noted that although sustainability has many elements, because of its power to combat poverty, scholars have focused mostly on financial sustainability. Financial sustainability is the capacity of MFI to use operating revenue to cover its financial and operational expenses.

Sekabira (2013), Quayes (2012), Ayayi and Sene (2010)

The regulations governing long-term sustainability need sound financial management, but the majority of government-backed microlending programs ignore this requirement as a prerequisite for sufficiency. This corroborates the claim made by Rhyne and Otero that MFIs require charitable contributions to continue operating profitably. It is recommended that MFIs investigate commercialization as a means of securing their financial viability and expanding their pool of potential investors. As debt has climbed and deposits have risen, there has been a discernible shift away from grants and soft loans, indicating a trend toward financial viability for MFI funding.

Kinde (2012)

They posited a connection between funding and the sustainability of the financial sector, noting that "different sources of capital could affect profitability, hence sustainability of MFIs." There is scant evidence since the relationship between MFI financing and financial sustainability is still unclear. Despite not explicitly examining the relationship, the research in this sector sheds light on how funding and MFI financial viability are perceived.

Atahau et al. (2021)

"Empowerment" initiatives are those that aim to increase capacity, particularly for populations that are more vulnerable. Following their empowerment, these marginalized groups are granted certain capacities, such as the ability to meet their most basic needs, experience independence, obtain resources that allow them to effectively increase their income to purchase the goods and services they require, participate in case-related decision-making, and promote development. Supporting and sustaining women-owned and -run small- and medium-sized businesses (SMEs) is the aim of the Women's Enterprise and Development Fund (WEDF). Women are still disproportionately underrepresented in entrepreneurship despite these efforts, which is why this study looked at the major variables affecting women's empowerment and entrepreneurship.

Eswarappa (2020)

The participation of women in community-based activities is demonstrated by this study. Organizations are crucial for implementing community development programs and empowering women. Microfinance increases the influence of low-income women. Microfinance has the capacity to significantly improve women's living conditions, promote gender parity in participation, and increase long-term revenue streams. Scholarly engagement is significantly hampered by the donors' involvement in people's daily life, particularly in case of women who attempt to sustain themselves. These issues encompass a variety of theoretical and empirical facets. This study examines the development initiatives implemented by community-based organizations (CBOs) in rural areas across the country.

Mwale et al. (2021)

The subject of this research is microfinance organizations, also referred to as joint and multiple duties. According to the study's scenario, group loans are regarded as common sense, especially in Javanese culture in Indonesia. One strategy employed by financial institutions to lower the risk of default on loans given to the community is group loans, since members share authority over payments and the use of the funds.

Tariq and Bhat (2020)

Women can invest in their microbusinesses and improve their standard of living because of microfinance. Giving low-income customers access to savings accounts, microloans, microinsurance, and other microfinance services is the fundamental idea of microfinance. Based on the data currently accessible, microfinance helps low-income women launch new microbusiness ventures and grow existing ones. Women have also been effective in generating jobs in their communities through their microbusiness endeavors. Women who take out loans also confront marketing issues related to modest loan amounts, fluctuating loan payback terms, and a dearth of information and assistance.

Masha (2021)

You can find a contribution by reading "The Role of Microfinance and Self-Help Groups in Women's Empowerment in Rural India." This study looks at the various ways that microfinance programs support women's empowerment. It is well known that microfinance, and especially microcredit, is one of the most crucial instruments for economic growth.

Self-help groups (SHGs), which essentially provide financial support to underprivileged people in rural areas, oversee the microfinance sector. Through these organizations, people can strengthen their capacity to assist themselves collectively, which will give them more power. The study looks at how microfinance helps women in rural India have more economic power and decrease poverty. Since microfinance increases women's confidence, drive, and skill development, it highlights the significance of microfinance for women's empowerment. Members of an SHG have a big say in whether or not women join.

5.5 CONCEPTUAL FRAMEWORK OF STUDY

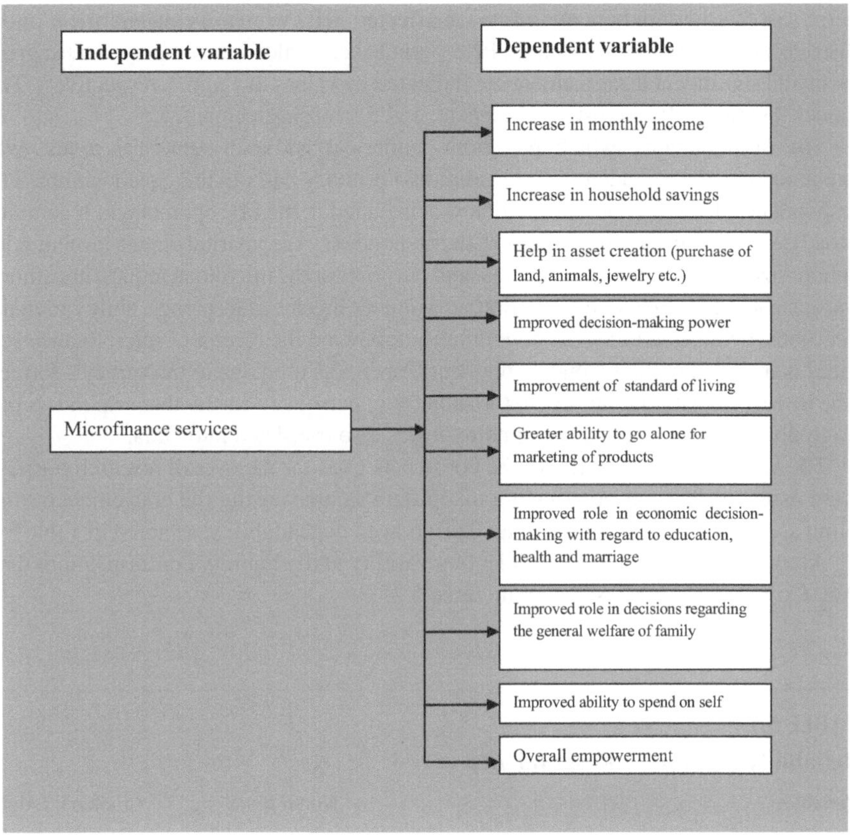

FIGURE 5.1 Figure illustrating the conceptual framework of the study.

5.6 METHODOLOGY

The survey conducted in the districts of Khurda and Sambalpur provided the fundamental data for the study. The purpose of this survey was to determine the effects of microfinance services on the mobility, asset ownership, income and saving, and decision-making of women in SHGs. A survey was issued and collected in the districts of Khurda and Sambalpur to gather the data. All of the data used to create the results came from people who answered the basic sample survey. Numerous statistical analyses and tests formed the basis of the study. The statistical package for social science (SPSS) program was used to assess the quantitative data. Based on the respondent profile, descriptive statistics were utilized to compute Cronbach's alpha and pivot tables.

The major survey for the study involved 960 SHG women from Khurda and Sambalpur districts; these women represented a variety of backgrounds and included married, single, widowed, and divorced women. Using structured questionnaires is one way to ensure the data is legitimate and trustworthy. The questionnaire for the study was developed based on a review of the literature. The primary goal of the questionnaire was to ascertain how microfinance affected SHG women's empowerment in the districts of Khurda and Sambalpur. Five-point Likert scales were employed. Disagreement and significant disagreement are indicated by values of 1 and 2, respectively. The values 3 indicate disinterest, 4 agreement, and 5 strong agreement.

The questionnaire, which had both numerical and non-numerical items, was explained to the respondents. There were two primary parts to the questionnaire. The respondent's demographic information was included in the first component. It contains broad details regarding the address of the respondent. Age, marital status, level of education, size of family, monthly income, and purpose for the microloan request are among the variables. Descriptive information on women's income and savings, ability to make decisions, ownership of assets, and mobility following the receipt of microfinance services is included in the second component. Paper was used due to the survey's format. The researcher gave the surveys to the subjects in person. To ensure that responders had no trouble completing the questionnaire, it was also translated into Odia.

The validity and reliability of the constructs used for the overall research purpose were assessed through pilot testing on 15 data samples using the convenience sampling approach; all values more than 0.60 showed dependability, as noted in Table 5.1.

After 960 respondents, data collection validity and reliability conforms more than 0.60 Cronbach's Alpha as shown in Table 5.2.

TABLE 5.1
Reliability Test for 15 Respondents

Variables	No. of Items	Cronbach's Alpha
Increase in monthly income	4	0.683
Improved decision-making power	5	0.774
Improvement of standard of living	4	0.785
Greater ability to go alone for marketing of products	4	0.682
Microfinance services	5	0.856

TABLE 5.2
Reliability Test for 960 Respondents

Variables	No. of Items	Cronbach's Alpha
Increase in monthly income	4	0.625
Improved decision-making power	5	0.623
Improvement of standard of living	4	0.743
Greater ability to go alone for marketing of products	4	0.714
Microfinance services	5	0.682

5.7 DATA ANALYSIS

This study focused on the influence of microfinance on women's entrepreneurship, specifically among SHG women in Sambalpur and Khurda districts. Based on the objectives mentioned earlier, the following hypotheses have been developed.

H1: Microfinance has significant impact on the monthly income of SHG women in Sambalpur and Khurda districts.
H2: Microfinance has significant impact on the decision-making power of SHG women in Sambalpur and Khurda districts.
H3: Microfinance has significant impact on the improvement of standard of living of SHG women in Sambalpur and Khurda districts.
H4: Microfinance has a significant impact on the ability to go alone for marketing of products by SHG women in Sambalpur and Khurda districts.

The replies of 660 respondents of Sambalpur district and 300 respondents from Khurda district were coded in SPSS after the instrument's reliability was established by pilot testing (Table 5.3).

Following the completion of the data entry, PIVOT tables are utilized to determine the characteristics of the respondents profile (Table 5.4).

TABLE 5.3
Descriptive Analysis

Increase in Monthly Income: The change in monthly income of respondents after participation in microfinance institution is analyzed based on count and percentage of participation on particular ordinal value.									
Cat-1		Cat-2		Cat-3		Cat-4		Total	
Count	Row N %	Count	Row N %	Count	Row N %	Count	Row N %	Count	Row N %
100	10.4%	200	20.8%	400	41.7%	260	27.1%	960	100.0%
Cat1-BaliantaBlock (Rural)									
				Khurda District					
Cat2-Bhubaneswar (Urban)									
Cat 3-JUJOMURA BLOCK									
				Sambalpur District					
Cat4-JAMANKIRA BLOCK									

TABLE 5.4
Increase in Monthly Income

Category	2		3		4		5		Total	
	Count	Row N %	Count	Row N %	Count	Row N %	Count	Row N %	Count	Row N %
Cat-1	0	0.0	2	2.0	43	43.0	55	55.0	100	100.0
Cat-2	0	0.0	2	1.0	89	44.5	109	54.5	200	100.0
Cat-3	14	3.5	64	16.0	150	37.5	172	43.0	400	100.0
Cat-4	0	0.0	0	0.0	128	49.2	132	50.8	260	100.0
Total	14	1.5	68	7.1	410	42.7	468	48.8	960	100.0

Notes: 1-Strongly Disagree, 2-Disagree, 3-Neutral, 4-Agree, 5-Strongly Agree

After joining a microfinance institution, 55% of the 100 members of Balianta Block in the Khurda area strongly feel that their monthly income has increased. Of the 200 members of the Khurda district in Bhubaneswar (Urban), 54.5% firmly concur that joining a microfinance organization increased their monthly income. Out of the 400 SHG members in Sambalpur district's Jujomura Block, 43% strongly agree that joining a microfinance institution increased their monthly income. Out of the 260 SHG members in the Jamankira Block of Sambalpur district, 50.8% strongly concur that joining a microfinance institution has increased their monthly income. After joining the microfinance institutions, 48.8% of the 960 members in the Sambalpur and Khurda districts strongly feel that their monthly income has increased.

5.7.1 Improved Decision-Making Power

Based on the count and percentage of involvement on a specific ordinal value, the enhanced decision-making ability of the respondents following their participation in microfinance organizations is assessed.

After joining a microfinance institution, 51% of the 100 members of the women's SHG in Balianta Block, Khurda district, concur that their ability to make decisions about their family's finances has improved. Out of 200 members of SHGs in Bhubaneswar (Urban), 58% of them concur that their ability to make decisions has improved since joining a microfinance institution. After joining microfinance institutions, 39% of 400 members of the women's SHG in Jujomura Block, Sambalpur district, concur that their ability to make decisions has improved. 46.9% of the 260 women's SHG members in the Jamankira Block of Sambalpur district concur that their ability to make decisions has improved since joining a microfinance institution. Of the 960 women's SHGs in Sambalpur and Khurda district, 46.7% concur that there is improvement in decision-making (Table 5.5).

TABLE 5.5
Improved Decision-Making Power (Roose & Shooemaker, 2002)

		Improved Decision-Making Power Regarding Economic Matters of Family									
		2		3		4		5		Total	
		Count	Row N %	Count	Row N %	Count	Row N %	Count	Row N %	Count	Row N %
Category	Cat-1	1	1.0	44	44.0	51	51.0	4	4.0	100	100.0
	Cat-2	2	1.0	52	26.0	116	58.0	30	15.0	200	100.0
	Cat-3	50	12.5	180	45.0	159	39.8	11	2.8	400	100.0
	Cat-4	32	12.3	105	40.4	122	46.9	1	.4	260	100.0
	Total	85	8.9	381	39.7	448	46.7	46	4.8	960	100.0

Notes: 1-Strongly Disagree, 2-Disagree, 3-Neutral, 4-Agree, 5-Strongly Agree

TABLE 5.6
Improvement in the Standard of Living (Roose & Shooemaker, 2002)

		Improvement of Standard of living									
		2		3		4		5		Total	
		Count	Row N %	Count	Row N %	Count	Row N %	Count	Row N %	Count	Row N %
Category	Cat-1	2	2.0	55	55.0	41	41.0	2	2.0	100	100.0
	Cat-2	3	1.5	72	36.0	99	49.5	26	13.0	200	100.0
	Cat-3	37	9.3	167	41.8	173	43.3	23	5.8	400	100.0
	Cat-4	53	20.4	121	46.5	84	32.3	2	.8	260	100.0
	Total	95	9.9	415	43.2	397	41.4	53	5.5	960	100.0

Notes: 1-Strongly Disagree, 2-Disagree, 3-Neutral, 4-Agree, 5-Strongly Agree

5.7.2 Improvement in Standard of Living

The improvement of the standard of living of the respondents after participating in microfinance institutions is analyzed based on the count and percentage of participation in particular ordinal values (Table 5.6).

After joining microfinance institutions, 41% of the 100 members of the women's SHGs in Balianta Block, Khurda district, concur that their standard of living has improved. Of the 200 women SHG members in Bhubaneswar (URBAN), 49.5% believe that their standard of life has increased as a result of using microfinance institutions. 43.3% of the 400 women's SHGs in Sambalpur district's Jujomura block concur that their standard of living has increased as a result of using microfinance

TABLE 5.7
Ability to go for Marketing Alone (Roose & Shooemaker, 2002)

		Greater Ability to go Alone for Marketing of Products by HR Intervention									
		2		3		4		5		Total	
		Count	Row N %	Count	Row N %	Count	Row N %	Count	Row N %	Count	Row N %
Category	Cat-1	12	12.0	56	56.0	31	31.0	1	1.0	100	100.0
	Cat-2	21	10.5	100	50.0	65	32.5	14	7.0	200	100.0
	Cat-3	11	2.8	131	32.8	233	58.3	25	6.3	400	100.0
	Cat-4	93	35.8	164	63.1	3	1.2	0	0.0	260	100.0
	Total	137	14.3	451	47.0	332	34.6	40	4.2	960	100.0

Notes: 1-Strongly Disagree, 2-Disagree, 3-Neutral, 4-Agree, 5-Strongly Agree

institutions. 32.3% of the 260 women's SHGs in Sambalpur district's Jamankira Block concur that their standard of living has increased as a result of joining microfinance institutions. In sum, 41.4% of the 960 members in the districts of Sambalpur and Khurda concur that there has been progress.

5.7.3 Ability to Go Alone for Marketing of Products by SHG Women

The ability to go alone for marketing of products after joining the microfinance institutions is analyzed by the count and percentage of participation on particular ordinal values (Table 5.7).

Regarding their capacity for marketing, 56.0% of the 100 members of the SHGs in Balianta Block of the Khurda district remain neutral. Out of the 200 members of the SHGs in Bhubaneswar (URBAN), 50% are agnostic about their capacity to market independently. Of the 400 members of Sambalpur district's Jujomura Block, 58.3% concur that joining microfinance institutions improves one's capacity to market goods on one's own through HR assistance. After joining a microfinance institution, 63% of the 260 female SHG members in the Jamankira Block of Sambalpur district remain undecided about their abilities to proceed independently with marketing. Out of the 960 SHG members in the districts of Khurda and Sambalpur, 47% are undecided about their abilities to market their products on their own without HR supervision. 34.6% of the 960 female members of the SHG, whereas remaining did not respond to the survey or partially responded with vague details.

5.7.4 Overall Empowerment

Of the 100 members of the SHGs in Balianta Block, 51% strongly concur that overall empowerment occurs. Of the 200 members of the SHGs in Bhubaneswar (Urban), 48.5% strongly concur that there is general empowerment. Out of the 400 women in

TABLE 5.8
Analysis – Empowerment

		Overall Empowerment									
		2		3		4		5		Total	
		Count	Row N %	Count	Row N %	Count	Row N %	Count	Row N %	Count	Row N %
Category	Cat-1	0	0.0	2	2.0	47	47.0	51	51.0	100	100.0
	Cat-2	0	0.0	20	10.0	83	41.5	97	48.5	200	100.0
	Cat-3	3	.8	87	21.8	225	56.3	85	21.3	400	100.0
	Cat-4	14	5.4	120	46.2	116	44.6	10	3.8	260	100.0
	Total	17	1.8	229	23.9	471	49.1	243	25.3	960	100.0

Notes: 1-Strongly Disagree, 2-Disagree, 3-Neutral, 4-Agree, 5-Strongly Agree

the Jujomura block's SHGs, 56.3% concur that general empowerment occurs (Table 5.8). Out of the 260 SHG members, 44.6% concur that general empowerment occurs. Of the 960 members of SHGs, 49.1% concur that overall empowerment occurs. After becoming a member of the microfinance institutions, the percentage change shown here happened. Microfinance seeks to empower women by giving them access to capital, training, and business opportunities. For various reasons, it is imperative to assess how microfinance affects women's empowerment. The empowerment of people is the first and most crucial component of sustainable development.

Therefore, it is critical to evaluate the impact of microfinance on women's empowerment and determine the validity of this assertion. According to Havbarhe and Izekor, receiving microcredit from a microfinance company has increased women's earnings and savings.

The study's primary findings support the assertion stated by Akter and colleagues.

It was discovered that every study hypothesis had strong validation. They argued that microcredit positively affects several aspects of life, such as a greater standard of living, improved decision-making abilities, a higher monthly income, the ability to do independent marketing, and overt empowerment. Chaudhury came to the conclusion that women's ownership rights over family property have slightly increased as a result of their involvement in microfinance programs. The poll also showed that microfinance has helped SHG members financial circumstances.

5.8 CONCLUSION

The expansion of the economic, social, and environmental pillars of sustainable rural development is significantly influenced by the resolution of resource restrictions that the rural populace faces. The availability of financial resources is one of the primary resource constraints that people living in rural areas must contend with. Microfinance originated in Bangladesh and can assist persons experiencing financial hardship in rural areas. Microfinance may help eradicate poverty, create jobs, empower women, ensure food security, improve health and education, and help prepare for climate change, according to a number of studies. Therefore, to accomplish sustainable rural development, microfinance can be an instrument for these communities' social, economic, and environmental advancement. In Odisha, microfinance is growing in popularity, especially among women who adhere to it.

Numerous studies have been conducted regarding the impact of microfinance on women's empowerment. The results demonstrate how microfinance can significantly improve women's empowerment in terms of income and savings, decision-making, and mobility. In addition to offering self-employment training as an HR solution, associated services such as training programs and seminars on microcredit are vital for women's empowerment since they provide them the capacity and self-assurance to make better decisions. Research indicates that microfinance may be a useful tool for empowering women because of the beneficial impacts it has had on women's social and economic situations. Microfinance programs in the districts of Sambalpur and Khurda have a major effect on SHG women. Establishing microfinance institutions is essential to enabling women to transform the economy.

REFERENCES

Agarwal M. (2016). Role of microfinance in empowering Indian women. *International Conference on Research in Humanities, Sociology and Education*, 5, 26–30.

Alshebami A. S. (2015). The role of microfinance for empowerment of poor women in Yemen. *International Journal of Social Work*, 2(1), 15–23.

Atahau A. D. R., Sakti I., Huruta A. D., Kim M.-S. (2021). Gender and renewable energy integration: The mediating role of green-microfinance. *Journal of Cleaner Production*, 318, 128536.

Ayayi A. G., Sene M. (2010). What drives microfinance's financial sustainability. *The Journal of Developing Areas*, 44, 303–324.

Basu J. P. (2006). Microfinance and women empowerment: An empirical study with special reference to West Bengal. *Journal of International Development*, 40(17), 66–72.

Borbora S., Sarma G. K. (2007). Microfinance institutions: Sustainability and outreach. *Indian Institute of Technology Guwahati working paper*.

Brau J., Woller G. (2004). Microfinance: A comprehensive review of existing literature. *Journal of Entrepreneurial Finance*, 9, 1–27.

Eswarappa K. (2020). Community-based organizations (CBOs) and their role in the development of women: A case from Andhra Pradesh. *Contemporary Voice of Dalit*, 13, 34–49.

Galiè A., Teufel N., Korir L., Baltenweck I., Girard A. W., Dominguez-Salas P., Yount K. M. (2019). The women's empowerment in livestock index. *Social Indicators Research*, 142(2), 799–825.

Iezza P. (2010). Financial sustainability of microfinance. Master's Thesis, Copenhagen Business School, Copenhagen.

Janda K., Turbat B. (2013). Determinants of the financial performance of microfinance institutions in central Asia. *Post Communist Economics*, 25, 557–568. https://doi.org/10.1080/14631377.2013.844935

Kabeer N. (2005). Gender equality and women's empowerment: A critical analysis of the third millennium development goal 1. *Gender & Development*, 13(1), 13–24.

Kar A. K. (2012). Does capital and financing structure have any relevance to the performance of microfinance institution? *International Review of Applied Economics*, 26, 329–348.

Kimando M. J., Kihoroj M., Njogu G.W. (2012). Factors influencing the sustainability of microfinance institutions in Murang'a municipality. *International Journal of Business and Commerce*, 1, 21–45.

Kinde B. A. (2012). Financial sustainability of microfinance Ethiopia. *European Journal of Business and Management*, 4.

Kipesha E. F., Zhang X. (2013). Sustainability, profitability and outreach tradeoffs: Evidence from microfinance institutions in East Africa. *European Journal of Business and Management*, 5, 136–148.

Masha S. (2021). Role of micro finance and self-help groups in women empowerment in Rural India. *International Journal of Management Reviews*, 9(2), 473–487.

Mwale M. L., Chirwa G. C., Mchenga M., Zabula T. K. (2021). Micro-finance and women's perception of domestic violence in a fragile state. *World Development Perspectives*, 24, 100374. https://doi.org/10.1016/J.WDP.2021.100374

Paul M. S. (2010). Bridging the gap to the microfinance promise: A proposal for a tax exempt microfinance hybrid entity. *International Law and Politics*, 42, 1383–1426. http://nyujilp.org/wpcontent/uploads/2010/10/nyi409-paul1.pdf

Postelnicu L., Hermes N. (2018). Microfinance performance and social capital: A cross-country analysis. *Journal of Business Ethics*, 153(2), 427–445.

Quayes M. M. (2012). Depth of outreach and financial sustainability of microfinance institutions. *Applied Economics*, 44, 3421–3433.

Rahman S., Junankar P. N., Mallik G. (2009). Factors influencing women's empowerment on microcredit borrowers: A case study in Bangladesh. *Journal of the Asia Pacific Economy*, 14(3), 287–303.

Raphael G., Mrema G. L. (2017). Assessing the role of microfinance on women empowerment: A case of PRIDE (T)Shinyanga. *Business and Economics Research*, 7(2), 35–54.

Sekabira H. (2013). Capital structure and its role on performance of microfinance institutions: The Ugandan case. *Sustainable Agriculture Research*, 2, 86–101.

Tariq S., Abbas Bhat M. (2020). Microfinance and women entrepreneurship: An extant literature review. *Journal of Entrepreneurship and Management*, 9, 18.

6 Does Personal Innovativeness Influence the Technology Adoption Behavior of Women Entrepreneurs?

A Positivist Study

Kharabela Rout, Priti Ranjan Sahoo, Mousumi Panda and Nik Zulkarnaen Khidzir

6.1 INTRODUCTION

In India, women entrepreneurs belong to the MSMEs which are considered a crucial growth engine for Indian economic development (Prabha et al., 2025; Ramya et al., 2024). The women entrepreneurs are facing various challenges in business operations, supply chain management, and marketing (Aravamudhan et al., 2024). To tackle these business challenges, women entrepreneurs rely on e-commerce and e-platforms for the survival of their businesses and to overcome the challenges. In a post-pandemic environment, the Indian MSMEs claimed that digital initiative is the need of the hour to take the business to the following levels (Mishra et al., 2021).

Moreover, according to Jiang (2020), digital technologies empowered the response strategy to the COVID-19 pandemic in the short term and the trend and effective way for the small business to connect and sell their products to wide groups of customers. It will act as a technological foundation for the business organization currently and in the long run. Hence, it is important to understand how the entrepreneurs in MSMEs adopt digital technologies and the factors that play an important role in adopting digital technologies. In the extant literature, most researchers explained the digital technology adoption by entrepreneurs in various contexts (Dabas et al., 2021; Bhagat & Sambargi, 2019; Olsson & Bernhard, 2020). Very limited studies examined the factors that influence the entrepreneurs belonging to MSMEs to adopt digital technologies. As previously discussed, women entrepreneurs in MSMEs bear importance for the economy, and they are adopting new digital technologies; hence, it is necessary to understand the phenomenon of digital technology adoption, particularly by women entrepreneurs in MSMEs by

DOI: 10.1201/9781003508953-6

which policymakers and government can get an insight about digital technology adoption by women entrepreneurs which will further support the policymaking on digitization of MSMEs. The extant literature overlooked the digital technology adoption by women leading MSMEs in India which was accelerated during the pandemic and still is an important trend for the small businesses; hence this current research seeks to understand the digital technology adoption by women entrepreneurs. E-commerce has evolved into a technological medium enabling companies to serve customers, extend their presence into new territories, address external pressures, and minimize expenses (Martin & Matlay, 2003). These platforms function as digital marketplaces where transactions occur over the internet, operating under either B2B or B2C models (Ayawei et al., 2023). Small business proprietors can leverage these platforms to engage with a broad digital audience, facilitating revenue growth and customer retention while requiring minimal investment in promotion, delivery, inventory, and warehousing. In this context, this study proposed a central research question as follows:

RQ: What are the factors motivating Indian women entrepreneurs belonging to MSMEs to adopt e-commerce platforms?

The current research grounded on the TAM to answer the research questions and bridge the knowledge gap in the extant literature. The rest of the study is as follows: first, this study identifies the factors that influence women entrepreneurs to adopt e-commerce platform and develops research hypotheses through literature review. It empirically validates the hypotheses by collecting primary data from women entrepreneurs and offering theoretical and policy implications.

6.2 THEORETICAL BACKGROUND AND HYPOTHESIS DEVELOPMENT

Davis (1985) established the TAM, which scholars have widely used for the last two decades to explain the adoption of digital technologies by organizations and individuals. TAM was developed by relying on the theory of reasoned action (TRA), which was grounded by Fishbein and Ajzen (1977). TAM explains the actual behavior toward a technology influenced by attitude toward the technology, and attitude was influenced by two constructs, namely perceived ease of use (PEOU) and perceived usefulness (PU). Also, perceived usefulness has a direct impact on actual behavior. Apart from TAM, in the extant literature, personal innovativeness (PI) is considered as an important factor that can influence the technology adoption of an individual (Lu et al., 2005).

6.2.1 PEOU

PEOU refers to the extent to which an individual perceives that utilizing a specific information system would entail minimal effort or be effortless (Davis, 1989) (Rahmayanti et al., 2021). Past research based on TAM stated that PEOU influenced the adoption intention of individuals for various technologies (Venkatesh & Davis, 2000; Gefen et al., 2003). Women entrepreneurs adopted e-commerce platforms due to strict lockdown and social distancing, so while adopting e-commerce platforms, they might be accessed that how easy and required fewer efforts to run and manage the e-commerce platforms by their employees and for them as well. In the extant literature, many studies have empirically confirmed that PEOU positively influences entrepreneurs' decision-making to adopt new technologies for their ventures (Oly Ndubisi et al., 2005; Rahman et al., 2017; Zaremohzzabieh et al., 2016). Moreover, Davis (1985) incorporated an important factor, an attitude toward technology, which the PEOU and PU can influence. In the context of entrepreneurs' digital technology adoption, the PEOU can positively influence the entrepreneurs' attitude toward a particular technology; in the same vein, PEOU can influence the women entrepreneurs' attitude toward e-commerce platforms.

Moreover, several studies confirm that PEOU has a positive effect on PU (Venkatesh & Davis, 2000; Gefen et al., 2003). PEOU has a direct effect on PU, and it is because, between the two constructs, a user will find the one which is easier to use as more useful (Davis, 1987). Hence the following hypotheses are advanced:

H_1: PEOU of e-commerce platform positively influences women entrepreneurs' adoption intention for e-commerce platforms.

H_2: PEOU positively influences the attitude of women entrepreneurs toward e-commerce platforms.

H_3: PEOU has a positive and significant impact on the PU of e-commerce platforms.

6.2.2 PU

Davis (1989) has defined PU as "the degree to which an individual believes that using a particular system would strengthen his or her job performance." A previous study by Roh and Park (2019) suggests that the intention to use a particular technology is related to its expected outcomes. In the premises of technology adoption, prior literature indicates that PU is a strong predictor of intention to adopt a technology (Taylor & Todd, 1995). PU of various technologies has a strong effect on the adoption intention of entrepreneurs (Chatterjee et al., 2021; Sentosa et al., 2011). PU also positively influences the attitude toward technology (Chawla & Joshi, 2019). In the same vein, the women entrepreneurs' attitude toward e-commerce platforms and adoption decisions can be influenced by the perceived usefulness or outcomes of the e-commerce platforms. Therefore, the following hypotheses are proposed:

H_4: PU has a positive and significant impact on women entrepreneur's adoption intention of e-commerce platforms.

H_5: PU has a positive and significant impact on the attitude of women entrepreneurs toward e-commerce platforms.

6.2.3 Attitude

An individual's attitude can be described as the level of evaluative sentiment they hold toward specific behaviors (Fishbein & Ajzen, 1975). In the theory of planned behavior, attitude is positioned as a direct precursor to behavioral intention (Ajzen, 1991). In TAM, Davis (1989) argued that a positive and negative attitude toward a particular technology can influence the adoption intention of the individual, and additionally, various studies have confirmed that attitude can influence the adoption intention (Weng et al., 2018; Wang et al., 2019; Hussein, 2017). In line with this, the attitude of women entrepreneurs toward e-commerce platforms can lead to the adoption of e-commerce platforms. Hence the following hypothesis is advanced as follows:

> H_6: Positive attitude of women entrepreneurs toward e-commerce platforms positively influences the adoption intention of e-commerce platforms.

6.2.4 Personal Innovativeness

PI can be defined as the "willingness of an individual to try out a new technology (Agarwal & Prasad, 1998; Patel, 2016). PI is conceptualized as a trait, i.e., relatively stable characteristics of individuals that are stable across situational consideration. Agarwal and Prasad (1998) argued that PI manifests in the technology acceptance behavior through its relationship with beliefs and perception. Individuals with higher PI develop a positive perception of a certain event. Prior literature in the context of technology acceptance behavior has examined the role of PI as antecedents and moderators (Lu et al., 2005; Lu, 2014). Bhagat and Sambargi (2019) study argued that PI of women entrepreneur leads to adoption of digital technologies. Consistent with Jackson et al. (2013) and Agarwal and Prasad (1998), in this study, we considered PI in e-commerce platforms of women entrepreneurs acts as an important factor that moderates the perception or attitude of women entrepreneurs and adoption intention, which theorized that the women entrepreneur with high PI could develop a positive attitude which leads to a higher chance to adopt the e-commerce platform. Hence in line with the earlier discussion, this study advanced the following hypothesis:

> H_7: The attitude of women entrepreneurs and adoption intention toward e-commerce platforms are strengthened by the higher PI of women entrepreneurs.

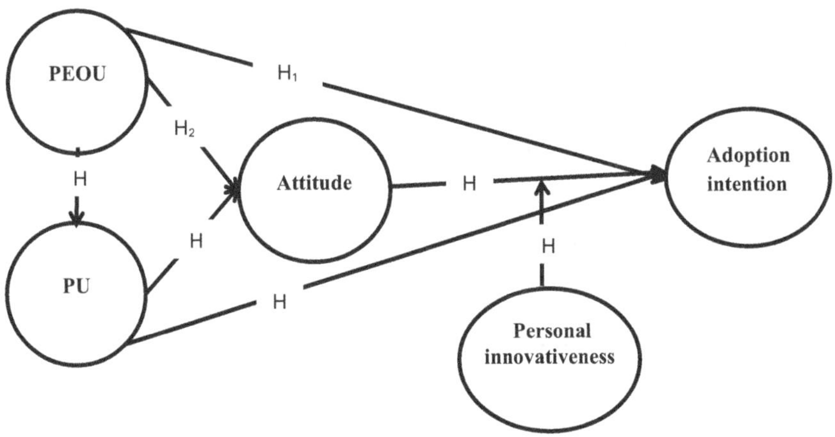

FIGURE 6.1 The conceptual model.

6.3 METHODS

To empirically validate and test the hypothesized model, this study followed a positivist approach and designed a survey to collect data from women entrepreneurs. In line with Alford and Page (2015), the convenience sampling technique was adopted in this study, where women entrepreneurs who are registered under the MSME department of the Government of India through the Udyoga Adhaar process and registered sellers in any B2C or B2B e-commerce platforms and operating in urban areas are included in the sample or considered as the sample unit of the study. In this study, we considered small business, and we define small business in line with the Government of India guidelines for MSMEs and the business which invested not more than 10 crores in Indian rupees to create the asset and machinery for the business and annual turnover not more than 50 crores in Indian rupees. A questionnaire was designed and distributed through online medium and out of the total 150 distributed questionnaires, 101 correctly filled questionnaires were considered for data analysis. The sample of women entrepreneurs falls into the age bracket of 23–45, out of which 40% are graduates, 36% have passed intermediate exams, and the rest 24% are postgraduates. This study considers samples from three industries: packaged food manufacturing, home care products, and apparel.

This study adopted scales from various studies to measure the latent variables illustrated in the conceptual framework. PEOU and PU scales with three items each were adopted from Hong et al. (2006). The scale for attitude with five things was adopted from Taylor and Todd (1995). AI is measured with three items adapted from Jaffa and Musa (2016). PI was adopted from Agarwal and Prasad (1998) with four items. All the scales are modified with the context of the study. The items are illustrated in Table 6.1.

This research utilized partial least squares structural equation modeling (PLS-SEM). As recommended by Asadi et al. (2017), the sample size for PLS-SEM should ideally be at least ten times greater than the number of independent variables in the model. Therefore, the sample size in this study is deemed sufficient. Smart PLS 4 software package was employed to conduct the PLS-SEM analyses.

6.4 DATA ANALYSIS

6.4.1 Measurement Model Assessment

During the initial phase of data analysis, an examination of the measurement model was conducted to evaluate its validity and reliability. Convergent validity was assessed by computing the average variance extracted (AVE), with a threshold of 0.5 recommended by Hair et al. (2011) to establish adequate convergent validity. The AVE values for each construct exceeded 0.5, indicating satisfactory convergent validity for the study. Subsequently, discriminant analysis was performed using the correlation matrix provided in Table 6.2. The diagonal elements represented the square roots of AVEs, which were higher than the correlation values off the diagonal. This outcome suggests that discriminant validity was achieved as outlined by Hair et al. (2011). The reliability of the constructs was evaluated by computing the composite reliability (CR) with a threshold of 0.7 suggested by Hair et al. (2011) for reliability. In this study, all constructs exhibited CR values exceeding 0.7, indicating reliability.

TABLE 6.1
Assessment of Measurement Models

Constructs (Items, AVE, CR)

PU (AVE: 0.778; CR:0.708)
PU1: Using an e-commerce platform helps me to do business quickly and smoothly.
PU2: Using an e-commerce platform makes our sales easier.
PU3: I find the e-commerce platform useful for my business.
PEOU (AVE: 0.648; CR:0.708)
PEU1: Learning how to manage sales and marketing in an e-commerce platform is easy.
PEU2: E-commerce platform is clear and understandable to use.
PEU3: I and most of my employees find the e-commerce platform easy to use.
Attitude (AVE:0.555; CR:0.801)
ATT1: Sales and marketing through e-commerce platforms offer real advantages for business over the physical form of business.
ATT2: I like using e-commerce platforms to do business through it.
ATT3: I think those who are small entrepreneurs should adopt e-commerce platforms for their business to expand.
ATT4: I think an e-commerce platform is an effective tool to do business effectively.
ATT5: I would like to suggest other entrepreneurs to adopt e-commerce platforms for their business.
PI (AVE: 0.653; CR:0.778)
PI1: If I become aware of a new information technology, I would seek opportunities to explore its functionality.
PI2: I tend to be the earliest adopter of new information technologies among my peers.
PI3: Typically, I am cautious when it comes to trying out unfamiliar information technologies.
PI4: I enjoy exploring and experimenting with novel information technologies.
AI (AVE: 0.888; CR:0.718)
AI1: I intend to continue to use e-commerce platform in the future for my business.
AI2: I expect that I would use an e-commerce platform in the future.
AI3: I plan to use an e-commerce platform in the near future.

TABLE 6.2
Fronell-Larcker Criterion (Discriminant Validity)

	1	2	3	4	5
PU (1)	0.882				
PEOU (2)	0.436	0.804			
ATTITUDE (3)	0.467	0.462	0.744		
PI (4)	0.287	0.215	0.209	0.808	
AI (5)	0.569	0.621	0.411	0.108	0.942

6.4.2 Inner Model Testing

All the proposed hypotheses were tested in this phase of data analysis. As explained in Table 6.3, except for H2, the rest of the hypotheses are supported, and all the hypotheses are significant at 0.05 significance level. PEOU of e-commerce platform positively impacts AI (β:0.498; p: 0.010), providing evidence for H_1. PEOU has a positive and significant effect on PU (β: 0.234, p: 0.020) which confirmed H_3. Out of all independent variables, PU has the most significant effect on AI (β:541, p:0.011), confirming the H_4. The relationship between PU and ATT is significant and positive (β: 0.386; p: 0.041), which supported H_5. The effect of ATT on AI was significant and positive (β:0.395; p:0.11), and all the independent variables positively influenced the AI. The coefficient of determination (R^2) values for AI and ATT were 0.775 and 0.543, respectively, indicating that 77% of the variance in adoption intention and 54% of the variance in attitude were explained by the model. For the moderation analysis, this study adopted a product indicator approach. The result of moderation analysis revealed that PI has a significant and positive moderation effect (β:0.211; p:0.043) on the relationship between attitude and AI, which confirmed the $H_{7.}$

6.5 DISCUSSION AND IMPLICATIONS

The aim of this study is to understand the digital technology adoption behaviour of women entrepreneurs from MSMEs. To address it, this study adopted TAM as a theoretical underpinning. In line with the model proposed by Agrawal and Prasad (1998), this study incorporated PI as a factor, which plays a crucial role in technology adoption. This study found that PI of women entrepreneurs moderates the relationship between attitude and adoption intention. Women entrepreneurs with higher PI and positive attitudes are more likely to adopt e-commerce platforms for their business. The PI is a trait developed by the women entrepreneurs because of the awareness and knowledge about the digital technologies. Second, the women entrepreneurs understand the importance of the digital technologies for their business and might be that leads to experiment with the digital technologies. Hence the PI enhances the positive attitude toward digital technologies among the entrepreneurs.

Further, the result of the study revealed that when women entrepreneurs perceived that the usefulness is high by adopting e-commerce platform as digital

TABLE 6.3
Hypothesis Testing Result

Hypothesized Paths	B	Significance Level	Outcome
H_1: PEOU -> AI	0.498	0.010	Accepted
H_2: PEOU -> ATT	0.116	0.589	**Rejected**
H_3: PEOU -> PU	0.234	0.020	Supported
H_4: PU -> AI	0.541	0.011	Supported
H_5: PU -> ATT	0.386	0.041	Supported
H_6: ATT -> AI	0.395	0.011	Supported
H_7: ATT*PI->AI	0.211	0.043	Supported

technology, then it leads to the strong intention of women entrepreneurs to adopt e-commerce platforms to do business; this insight is in line with the study by Chatterjee et al. (2021) and Davis (1989). Moreover, perceived as easy to use has no significant impact on AI, which can be summarized as the easy-to-use digital technology has no significant impact on the adoption intention of women entrepreneurs, which contradicts the findings of the previous studies by Venkatesh and Davis (2000); Gefen et al. (2003), which found a strong correlation between PEOU and AI. The outcomes of the study also revealed that PEOU has a strong and positive effect on PU, which is in line with the study by Davis et al. (1989), which indicates that when technology is easy to use, women entrepreneurs perceived it as a useful technology, for instance, e-commerce platforms are beneficial for their business to scale and connect with large customers. Both PEOU and PU were found to positively and significantly influence the attitude of women entrepreneurs toward e-commerce platform as a technology, which is consistent with the study by Davis (1987) when the digital technology is easy to use and useful for a particular job. A positive attitude is formed by the women entrepreneurs for the specific technology. The study's findings also confirmed that a positive attitude of women entrepreneurs toward the e-commerce platform leads to the adoption of e-commerce platforms which is in line with the study Ajzen (1991).

This study advanced a TAM to explain the AI of e-commerce platform among the women entrepreneurs, which contributes to the existing body of knowledge on digital technology adoption by women entrepreneurs. The previous study overlooked factors like PI to explain the phenomenon, and this study bridges the gap by incorporating PI as a moderator between ATT and AI. Moreover, this study established the relationship between PEOU and PU for the use of e-commerce platforms among the women entrepreneurs. This study also examined the impact of PU and PEoU on the attitude of women toward e-commerce platforms for their business. In the context of Indian women entrepreneurs, this study is established these novel relationships which enrich the knowledge on Indian women entrepreneurs' attitude and adoption intention toward e-commerce platforms.

This study might support the policymakers and other administrative bodies to understand the digital technology AI of women entrepreneurs and might help in policymaking in the post-pandemic era; also, it helps e-commerce platforms to understand the phenomenon and allow a sizeable onboard number of audiences by improving the feature and design of platforms which support women entrepreneurs.

6.6 LIMITATIONS AND FUTURE RESEARCH IMPLICATIONS

A few limitations are highlighted in this study. This study was carried out with a limited number of samples and many variables. This study was carried out within one state, which might be a limitation to generalize the result for other states of India. Future research can use this model as a base model to explain various entrepreneurs' digital technology adoption intention with larger sample sizes and varied geographical regions. The prospective study may consider multiple factors like leadership support, self-efficacy, and techno stress to understand this phenomenon holistically.

REFERENCES

Agarwal, R., and Prasad, J. (1998), "A conceptual and operational definition of personal innovativeness in the domain of information technology", *Information Systems Research*, INFORMS, Vol. 9 No. 2, pp. 204–215.

Ajzen, I. (1991), "The theory of planned behavior-organizational behavior and human decision processes 50", *Ajzen, I. (2002) Perceived Behavioural Control, Self-Efficacy, Locus of Control and the Theory of Planned Behaviour. Journal of Applied Social Psychology*, Vol. 32 No. 4, pp. 665–683.

Alford, P., and Page, S. J. (2015), "Marketing technology for adoption by small business", *The Service Industries Journal*, Taylor & Francis, Vol. 35 No. 11–12, pp. 655–669.

Aravamudhan, V., Sivakumar, K., Vishnu, C. R., and Mohanasundaram, K. (2024), "Challenges faced by women entrepreneurs in rural India: A mixed-method study on perception", *The International Journal of Entrepreneurship and Innovation*, 14657503241254954.

Asadi, S., Nilashi, M., and Yadegaridehkordi, E. (2017), "Customers perspectives on adoption of cloud computing in banking sector", *Information Technology and Management*, Springer, Vol. 18 No. 4, pp. 305–330.

Ayawei, M. J., Raborife, M., and Maduku, D. K. (2023), "Assessing the factors underlying the adoption of e-commerce among B2B SMEs: A two-country study", *Journal of Electronic Commerce in Organizations (JECO)*, Vol. 21 No. 1, pp. 1–26.

Bhagat, R., and Sambargi, S. (2019), "Evaluation of personal innovativeness and perceived expertise on digital marketing adoption by women entrepreneurs of micro and small enterprises", *International Journal of Research and Analytical Reviews (IJRAR)*, Vol. 6 No. 1, pp. 338–351.

Chatterjee, S., Chaudhuri, R., Vrontis, D., and Basile, G. (2021), "Digital transformation and entrepreneurship process in SMEs of India: A moderating role of adoption of AI-CRM capability and strategic planning", *Journal of Strategy and Management*, Vol. 15 No. 3, pp. 416–433.

Chawla, D., and Joshi, H. (2019), "Consumer attitude and intention to adopt mobile wallet in India–An empirical study", *Journal of Strategy and Management* Vol. 15 No. 3, pp. 416–433.

Dabas, S., Sharma, S., and Manaktola, K. (2021), "Adoption of digital marketing tools in independent businesses: Experiences of restaurant entrepreneurs in India and United Kingdom", *Worldwide Hospitality and Tourism Themes*, Vol. 13 No. 2, pp. 214–235. https://doi.org/10.1108/WHATT-09-2020-0120

Davis, F. (1989), "Perceived usefulness, perceived East of use, and user acceptance of information technology", *MIS Quarterly*, Vol. 13 No. 3, pp. 319–340.

Davis, F. D. (1985), "A technology acceptance model for empirically testing new end-user information systems: Theory and results", Massachusetts Institute of Technology.

Fishbein, M., and Ajzen, I. (1977), "Belief, attitude, intention, and behavior: An introduction to theory and research", Addison-Wesley.

Gefen, D., Karahanna, E., and Straub, D. W. (2003), "Trust and TAM in online shopping: An integrated model", *MIS Quarterly*, JSTOR, pp. 51–90.

Hair, J. F., Ringle, C. M., and Sarstedt, M. (2011), "PLS-SEM: Indeed a silver bullet", *The Journal of Marketing Theory and Practice*, Vol. 19 No. 2, pp. 139–152.

Hong, S., Thong, J. Y. L., and Tam, K. Y. (2006), "Understanding continued information technology usage behavior: A comparison of three models in the context of mobile internet", *Decision Support Systems*, Elsevier, Vol. 42 No. 3, pp. 1819–1834.

Hussein, Z. (2017), "Leading to intention: The role of attitude in relation to technology acceptance model in e-learning", *Procedia Computer Science*, Elsevier, Vol. 105, pp. 159–164.

Jackson, J. D., Mun, Y. Y., and Park, J. S. (2013), "An empirical test of three mediation models for the relationship between personal innovativeness and user acceptance of technology", *Information & Management*, Elsevier, Vol. 50 No. 4, pp. 154–161.

Jaffar, M. A., and Musa, R. (2016), "Determinants of attitude and intention towards Islamic financing adoption among non-users", *Procedia Economics and Finance*, Elsevier, Vol. 37, pp. 227–233.

Jiang, X. (2020), "Digital economy in the post-pandemic era", *Journal of Chinese Economic and Business Studies*, Taylor & Francis, Vol. 18 No. 4, pp. 333–339.

Lu, J. (2014), "Are personal innovativeness and social influence critical to continue with mobile commerce?", *Internet Research*, Emerald Group Publishing Limited., Vol. 24 No. 2, pp. 134–159.

Lu, J., Yao, J. E., and Yu, C.-S. (2005), "Personal innovativeness, social influences and adoption of wireless internet services via mobile technology", *The Journal of Strategic Information Systems*, Elsevier, Vol. 14 No. 3, pp. 245–268.

Martin, L. M., and Matlay, H. (2003). Innovative use of the internet in established small firms: The impact of knowledge management and organisational learning in accessing new opportunities. *Qualitative Market Research: An International Journal*, Vol. 6 No. 1, pp. 18–26.

Mishra, M. N., Mohan, A., and Sarkar, A. (2021), "Role of lean six sigma in the Indian MSMEs during COVID-19", *International Journal of Lean Six Sigma*, Vol. 12 No. 2, pp. 262–289.

Olsson, A. K., and Bernhard, I. (2020), "Keeping up the pace of digitalization in small businesses–Women entrepreneurs' knowledge and use of social media", *International Journal of Entrepreneurial Behavior & Research*, Vol. 27 No. 2, pp. 378–396. https://doi.org/10.1108/IJEBR-10-2019-0615

Oly Ndubisi, N., Gupta, O. K., and Ndubisi, G. C. (2005), "The moguls' model of computing: Integrating the moderating impact of users' persona into the technology acceptance model", *Journal of Global Information Technology Management*, Taylor & Francis, Vol. 8 No. 1, pp. 27–47.

Patel, V. (2016), "Use of mobile wallet service by the youth: A study based in Ahmedabad.", *ASBM Journal of Management*, Vol. 9 No. 2.

Prabha, S. A., Palanichamy, N. V., Murugananthi, D., Shivakumar, K. M., Kalpana, M., and Balakrishnan, M. (2025). Research trends and dynamics in women entrepreneurship: A comprehensive bibliometric review. *Future Business Journal*, Vol. 11 No. 1, pp. 1–14.

Rahman, S. A., Taghizadeh, S. K., Ramayah, T., and Alam, M. M. D. (2017), "Technology acceptance among micro-entrepreneurs in marginalized social strata: The case of social innovation in Bangladesh", *Technological Forecasting and Social Change*, Elsevier, Vol. 118, pp. 236–245.

Rahmayanti, P., et al. (2021), "Integration of technology acceptance model and theory of reasoned action in predicting e-wallet continuous usage intentions", *International Journal of Data and Network Science*, Vol. 5 No. 4, pp. 649–658.

Ramya, U., Pushpa, A., and Ghosh, N. (2024). Women entrepreneurship–A way towards sustainability. In *The framework for resilient industry: A holistic approach for developing economies* (pp. 281–299). Emerald Publishing Limited.

Roh, M., and Park, K. (2019), "Adoption of O2O food delivery services in South Korea: The moderating role of moral obligation in meal preparation", *International Journal of Information Management*, Elsevier, Vol. 47, pp. 262–273.

Sentosa, I., Nejatian, H., Piaralal, S. K., and Faisal, A. (2011), "Empirical study of internet adoption among small And medium entrepreneurs (SMEs) in Malaysia (A structural equation modeling approach)", *2nd International Conference on Business and Economic Research (ICBER)*.

Taylor, S., and Todd, P. A. (1995), "Understanding information technology usage: A test of competing models", *Information Systems Research*, INFORMS, Vol. 6 No. 2, pp. 144–176.

Venkatesh, V., and Davis, F. D. (2000), "A theoretical extension of the technology acceptance model: Four longitudinal field studies", *Management Science*, INFORMS, Vol. 46 No. 2, pp. 186–204.

Wang, Y., Jin, L., and Mao, H. (2019), "Farmer cooperatives' intention to adopt agricultural information technology—Mediating effects of attitude", *Information Systems Frontiers*, Springer, Vol. 21 No. 3, pp. 565–580.

Weng, F., Yang, R.-J., Ho, H.-J., and Su, H.-M. (2018), "A TAM-based study of the attitude towards use intention of multimedia among school teachers", *Applied System Innovation*, Multidisciplinary Digital Publishing Institute, Vol. 1 No. 3, p. 36.

Zaremohzzabieh, Z., Samah, B. A., Muhammad, M., Omar, S. Z., Bolong, J., Hassan, S. B. H., and Mohamed Shaffril, H. A. (2016), "Information and communications technology acceptance by youth entrepreneurs in rural Malaysian communities: The mediating effects of attitude and entrepreneurial intention", *Information Technology for Development*, Taylor & Francis, Vol. 22 No. 4, pp. 606–629.

7 The Influence of Diversity in Women's Entrepreneurship on the Efficacy of Industry 4.0 Initiatives

Saumya Agarwal and C.K. Tiwari

7.1 INTRODUCTION

Industry 4.0 is ushering in a new era of global industrial transformation through the integration of digital technologies such as artificial intelligence (AI), the Internet of Things (IoT), big data analytics, and automation. In this context, diversity, particularly through women entrepreneurs, plays a vital role in shaping the success of these initiatives. Women bring distinct perspectives and problem-solving approaches that have been instrumental in driving innovation within Industry 4.0. This chapter delves into how women entrepreneurs contribute to the development and implementation of Industry 4.0 and how their diversity positively impacts the effectiveness of these initiatives.

7.2 WOMEN'S ENTREPRENEURSHIP

Hisrich and Brush (1984) defined women's entrepreneurship as the process of *creating, organizing, and managing a business* while facing challenges, with a particular focus on how gender affects the ability to innovate and establish businesses in traditionally male-dominated fields.

> In their research, Morris et al. (2006) describe *women entrepreneurs* as women who *innovate, initiate, and lead ventures* that aim to create and manage new enterprises, thus contributing to economic development.

Women's entrepreneurship in today's era is all about breaking barriers, fostering innovation, and promoting inclusivity. Women entrepreneurs are increasingly leading the way in various industries, especially in technology and digital spaces, where their creativity and unique perspectives are driving change. They are focused not

DOI: 10.1201/9781003508953-7

only on business success but also on addressing societal challenges, promoting sustainability, and creating opportunities for other women.

Today, the key aspects of women's entrepreneurship include:

1. *Innovation and Disruption:* Women entrepreneurs are at the forefront of innovation, often introducing new ideas, products, and business models that challenge traditional industries and contribute to economic growth.
2. *Tech and Digital Transformation:* More women are venturing into tech-driven fields, including AI, IoT, and fintech, bridging the gender gap in sectors once dominated by men. They are leveraging digital tools to scale businesses and enhance efficiency.
3. *Social Impact:* Many women entrepreneurs are deeply committed to social causes, using their businesses as platforms to address issues like environmental sustainability, gender equality, and community development.
4. *Leadership and Empowerment:* Women are taking leadership roles, not just in running businesses but in shaping policies, mentoring other women, and influencing industries to be more inclusive and diverse.
5. *Networking and Collaboration:* The rise of supportive networks and collaborations among women entrepreneurs is crucial for growth. Through these networks, they share knowledge, access funding, and empower each other to overcome challenges.
6. *Work–Life Balance:* Today, women entrepreneurs are also navigating the challenges of balancing business, family, and personal life, often adopting flexible business models that support work–life harmony.

Diversity in entrepreneurship refers to the inclusion of individuals from various demographic backgrounds, including gender, ethnicity, and socioeconomic status, in entrepreneurial roles. In the context of women's entrepreneurship, diversity encompasses a range of leadership styles, experiences, and ideas. Although women remain underrepresented in tech-driven sectors, increasing recognition of gender equality, along with advancements in technology and societal change, has led more women to assume leadership positions, driving industry innovations and transformations.

Gardner and Laskin (2017) defined diversity as: "differences among individuals based on various characteristics including race, gender, sexual orientation, socioeconomic status, and education." Their work focuses on the *business* and *organizational context*, underscoring the benefits of a diverse workforce, such as improved creativity and decision-making.

Hamilton (2005) defines diversity as "the presence of varied social and cultural characteristics within a group or society." Her definition is particularly relevant in the context of organizations and emphasizes that diversity includes race, ethnicity, gender, and other factors like age and disability.

Women entrepreneurs bring a unique approach to the market, offering distinct advantages in shaping and executing Industry 4.0 strategies. Their perspectives often disrupt conventional business models, leading to innovative, inclusive solutions. Additionally, businesses led by women are increasingly embracing socially responsible and sustainable practices, aligning with the core values of Industry 4.0, which emphasizes ethical technology and sustainability.

7.3 INDUSTRY 4.0 AND ITS TRANSFORMATIVE IMPACT ON BUSINESS

Industry 4.0 refers to the fourth industrial revolution, which is driven by the integration of digital technologies and smart systems into industrial processes. It encompasses the use of cyber-physical systems, the Internet of Things (IoT), artificial intelligence (AI), big data analytics, cloud computing, and automation to create a more interconnected, efficient, and intelligent manufacturing environment. Industry 4.0 marks a shift from traditional manufacturing methods to highly automated and data-driven production systems that enable real-time communication between machines, devices, and people.

World Economic Forum (WEF) often defines Industry 4.0 as *the fusion of advances in artificial intelligence, robotics, the Internet of Things, genetic engineering, quantum computing, and other technologies.*

The key features of Industry 4.0 include:

- *Automation:* The implementation of robotics, AI, and machine learning to automate production.
- *Big Data:* The use of extensive data analysis for informed decision-making.
- *Smart Technologies:* The deployment of IoT to link people, machines, and systems.
- *Customization:* The ability to offer personalized products and services using advanced technologies.
- *Cloud Computing:* Storing and accessing data and applications over the internet, allowing for better scalability and collaboration.

7.4 INDUSTRY 4.0 HOLDS IMMENSE IMPORTANCE IN ENHANCING ORGANIZATIONAL EFFICACY

1. **Increased Efficiency and Productivity**
   One of the most immediate benefits of Industry 4.0 is its ability to significantly *boost productivity* through automation and real-time data analytics.
2. **Enhanced Decision-Making through Real-Time Data**
   Industry 4.0 technologies, such as the *Internet of Things (IoT)* and *big data analytics*, enable real-time data collection and analysis.
3. **Customization and Personalization**
   In the past, mass production often meant limited options for customers. Industry 4.0 enables *mass customization*, allowing businesses to offer personalized products without compromising on efficiency or cost.
4. **Cost Reduction**
   Industry 4.0's automation, predictive maintenance, and enhanced supply chain management help businesses *reduce costs* in several ways:

a. *Lower labor costs*: Automation reduces the need for manual labor in many tasks.
b. *Reduced downtime:* Predictive maintenance ensures machines are operating at peak efficiency, minimizing unexpected breakdowns.
c. *Optimized resource usage:* Real-time data allows businesses to optimize raw material usage, energy consumption, and waste management.

5. **Sustainability and Environmental Benefits**
Industry 4.0 contributes to *sustainability* by enabling businesses to use resources more efficiently and reduce waste. For example:
a. *Energy-efficient processes:* Smart systems can optimize energy consumption, leading to lower carbon footprints.
b. *Waste reduction:* IoT devices and sensors can track waste in real time, allowing businesses to adjust processes to minimize material waste.

6. **Improved Collaboration and Connectivity**
Industry 4.0 fosters a *highly connected ecosystem*, where:
a. Machines, devices, and systems communicate seamlessly via IoT, creating more efficient workflows.
b. Supply chain partners, manufacturers, and even customers are better connected, allowing for smoother collaboration and faster product development.

As these technologies reshape industries, the role of women entrepreneurs becomes increasingly important. Their leadership in technology-driven sectors is crucial for the successful adoption and execution of Industry 4.0 strategies.

7.5 ROLE OF WOMEN ENTREPRENEURS IN SHAPING INDUSTRY 4.0

Women entrepreneurs are strategically positioned to influence the development and success of Industry 4.0 due to several key factors:

1. *Innovation Leadership:* Women entrepreneurs often challenge traditional business models, fostering innovation. Their problem-solving strategies are collaborative and interdisciplinary, which is particularly beneficial in the context of Industry 4.0. For instance, women-led AI and robotics startups are increasingly incorporating sustainability into their products, creating environmentally responsible technologies.
2. *Inclusive Technology Development:* Many women entrepreneurs focus on creating accessible technologies that cater to diverse populations. Their commitment to digital inclusion and gender equality in tech leads to the development of more user-friendly solutions, enhancing the effectiveness of Industry 4.0 initiatives.
3. *Networking and Empowerment:* Women entrepreneurs often build supportive, collaborative networks that encourage mentorship, knowledge sharing, and professional growth, especially in industries where women are historically underrepresented. These networks help women in tech expand their

careers and foster diversity within the sector, which is critical for the success of Industry 4.0.

4. *Cultural Transformation:* Women entrepreneurs prioritize creating inclusive, transparent, and supportive work environments, which boost employee satisfaction and productivity. These cultural changes contribute to the smooth integration of new technologies, as employees are more likely to embrace technological shifts in a positive work atmosphere.
5. *Sustainability Focus:* Many women entrepreneurs lead the charge in promoting sustainability in the context of Industry 4.0. Their focus on social impact, coupled with technological advancements, has resulted in sustainable business models that align with global sustainability goals, such as eco-friendly manufacturing and energy-efficient technologies.

7.6 THE EFFECTIVENESS OF INDUSTRY 4.0 WITH DIVERSE LEADERSHIP

Gender diversity in leadership is increasingly recognized as a crucial factor for the successful implementation of Industry 4.0 strategies. Research suggests that organizations with diverse leadership teams exhibit higher levels of creativity, innovation, and problem-solving capabilities. These attributes are essential for the effectiveness of Industry 4.0 initiatives because:

1. *Diverse Perspectives Enhance Solutions:* Women entrepreneurs bring different viewpoints, leading to more creative and adaptable solutions for Industry 4.0 challenges. This diversity of thought is vital in an ever-evolving technological landscape.
2. *Increased Innovation:* Diverse teams are better at identifying new opportunities and developing innovative solutions. In the context of Industry 4.0, this could involve the creation of new technologies, business models, and management systems for data and automation.
3. *Improved Performance:* A diverse leadership team often results in better decision- making and overall performance. Women entrepreneurs tend to foster collaboration, which is essential for driving the success of Industry 4.0 transformations.
4. *Broader Consumer Insights:* The involvement of women in leadership positions ensures that the needs of women and other underrepresented groups are considered in product and service design. This creates offerings that are more inclusive, aligning with the broader consumer base and contributing to the success of Industry 4.0 ventures.

7.7 CHALLENGES FACED BY WOMEN ENTREPRENEURS IN PROMOTING EFFICACY OF INDUSTRY 4.0

Despite the significant advantages of women's entrepreneurship in shaping Industry 4.0, challenges remain:

1. *Access to Funding:* Women entrepreneurs, particularly in technology sectors, often face difficulties securing funding due to gender biases among investors, which limits their ability to scale and invest in cutting-edge technologies.
2. *Gender Bias and Stereotypes:* In the male-dominated tech industry, women often encounter unconscious biases and stereotypes that challenge their authority and credibility. Addressing these biases is crucial for creating inclusive environments.
3. *Lack of Mentorship and Networking:* The absence of mentorship and professional networks for women in tech limits their access to valuable industry knowledge, resources, and opportunities.
4. *Work–Life Balance:* Women entrepreneurs, especially in tech-driven fields requiring high levels of expertise, often struggle to balance business demands with family responsibilities, hindering their full engagement with Industry 4.0 initiatives.

7.8 CONCLUSION

Women entrepreneurs play an essential role in enhancing the success of Industry 4.0 initiatives thereby promoting organizational efficacy. Their diverse leadership, innovative strategies, and commitment to inclusivity significantly impact the adoption and implementation of digital technologies across industries. As more women assume leadership roles in tech-driven sectors, their influence will continue to shape the future of Industry 4.0.

However, for women's entrepreneurship to reach its full potential, it is necessary to address systemic challenges such as access to capital, mentorship, and overcoming gender biases. By cultivating an inclusive environment that embraces diversity and innovation, Industry 4.0 initiatives can benefit from a wide range of perspectives, leading to more effective, sustainable, and equitable technological advancements.

REFERENCES

Gardner, J. E., & Laskin, A. (2017). Diversity in the Workforce: A Review of the Literature and Implications for Business and Organizational Practice. *Journal of Business Diversity*, 17(2), 58–71.

Hamilton, R. S. (2005). Diversity in the Workplace: The Role of Race, Gender, and Cultural Characteristics. In *Workplace Diversity and Inclusion* (pp. 101–118). Harvard University Press.

Hisrich, R. D., & Brush, C. G. (1984). The Woman Entrepreneur: A Comparative Analysis. In *Frontiers of Entrepreneurship Research* (pp. 315–338). Wellesley, MA: Babson College.

Morris, M., Miyasaki, N., Watters, C., & Coombes, S. (2006). The Dilemma of Growth: Understanding the Context and Process of Women Entrepreneurs' Business Expansion. *Journal of Business Venturing*, 21(2), 232–257. https://doi.org/10.1016/j.jbusvent.2005.04.003

8 Gender Parity in Entrepreneurship

A Dream Versus Reality in African Countries

Sheetal Chadda, Yukti Ahuja, and Franklin Easter Dumbari

8.1 INTRODUCTION

Women's entrepreneurship has been a subject of scholarly investigation in recent times, mainly due to the recognition that females need to be part of the mainstream workforce or labour workforce. While the investigation of corporate readiness for diversity and inclusivity is one area of study, the inclusivity of women in the entrepreneurial ecosystem is another. The entrepreneurship ecosystem has seen rapid growth in terms of the availability of finance, education, and mentoring; yet women's entrepreneurship has not taken off fully. As per a study by the World Bank, most women in Africa start their ventures due to necessity and not because they are able to identify the opportunities in the ecosystem. In addition, women traditionally tend to be in less profitable sectors. For ages, gender has been associated with work and roles. It is only the mothers who are required to take the larger responsibility of child-rearing in the households and care work for the elderly. At work, as well, many roles like front office manager are typically seen as female job profiles, while marketing managers are usually males. Likewise, higher job positions are typically attributed to males, resulting in the glass ceilings for females. In entrepreneurship, women also tend to face glass walls, consequent to which less female activity is noticed in the male-dominated sectors (MDS). Underlying prejudices, societal norms, insufficient exposure to the industry and business, and limits of time and capital are among the several issues impeding women's entry into male-dominated fields. Overall, Africa as a continent has been positive on entrepreneurship, and 76% of Africans view entrepreneurship as a beneficial career choice. This is the highest rate in the world. Africa is largely less developed, so it is deprived of many infrastructural facilities; yet in research, it was found that economics students from a university were more interested in entrepreneurship than their counterparts from the European Union. They showed higher entrepreneurial intentions as opposed to European Union students (Skrbková & Rydvalová, 2023).

DOI: 10.1201/9781003508953-8

8.2 LITERATURE REVIEW

In recent times, the academic literature has grown tremendously, and a great focus is also laid on the gender-related aspects of entrepreneurship (Panda, 2018). Women's entrepreneurship in developing and less developed countries is trending due to a lack of adequate knowledge in the same. Entrepreneurship is a growth engine for economies, and researchers started publishing studies on female entrepreneurship in academic journals over many decades, but a meagre 3.4% of the articles were published between 1987 and 1996, while 8.2% were published between 1997 and 2006 (Ojong et al., 2021). Due to widespread interest in the world's academia in the status of women's entrepreneurship, a spurt of rise is evident. Multiple researchers have identified women's entrepreneurship as necessity-driven and inspired by the concept of Western philosophy's empowerment of women. Not much is known about the African history and background, resulting in a metaphorical understanding of the drivers and the challenges of the entrepreneurial activity in the country. As per the World Bank's estimates, many women are indulging in entrepreneurial activities across Africa, running high-tech firms as CEOs to running 'mama mbagos', women selling vegetables, and an estimated 24% are running their ventures. A quarter of businesses are run by women, as opposed to only 5.7% of women's entrepreneurial ventures in European countries. Not only are women increasingly assuming entrepreneurial roles, but they are also assuming crucial leadership positions in the economy. In 2018, women in sub-Saharan Africa held 24.3% of board seats in African companies and represented 24.5% of board chairs, compared to the global average of 27% on corporate boards. Since 2010, women's engagement in the industrial and service sectors has risen, while in agriculture, women represent 40% of the workforce in crop production. The reason for growth is due to many initiatives like 'She Wins Africa' by the International Finance Corporation (IFC), AFAWA (Affirmative Finance Action for Women in Africa), a pan-African initiative by the African Development Bank, and the like. While these initiatives are fuelling growth, not all the pictures are rosy, and there are a multitude of factors that are the challenges as well. The following sections deal with factors that are posing challenges to women entrepreneurs.

8.2.1 Unpaid Care Work

Women tend to attend to multiple unpaid work at home, like care work for kids and the elderly; this work is neither considered as economic output nor paid. Estimates suggest that if women's unpaid care work were monetised, it would represent between 10% and 39% of GDP, perhaps surpassing contributions from the manufacturing, commerce, or transportation sectors in any specific country. However, this work takes lots of their time and effort; consequentially, the window for economic productivity is less, both temporally and physically. This is one reason women cannot contribute to the economy, which affects their social status. It is important to streamline the traditional gender roles; otherwise, most nations will not be able to reap the economic benefits of their women's potential. Despite the constraints of low resources and stringent economic conditions in most African economies, intentional measures can be implemented to tackle unpaid care responsibilities. Such actions will result in the benefits of mitigating gender disparity, fostering job creation, enhancing economic productivity, and generating chances for income advancement.

8.2.2 Access to Funds

Globally, women's access to financial resources is significantly limited. Notwithstanding significant general advancement—in 2017, the World Bank stated that 1.2 billion more individuals possessed bank accounts compared to 2011—there is a 9% disparity in access between women and men. In sub-Saharan Africa, about 37% of women possess a bank account, in contrast to 48% of men, a disparity that has increasingly expanded in recent years. Despite the existence of several women-orientated schemes like She Wins Africa and other micro-lending platforms, the limited access to the funds is largely due to women's inability to provide collateral, a lack of understanding of the norms, and other factors like financial issues are largely the domain of males in the family. Even in the event of access, the terms of conditions of loans or other financial supports are more stringent for females as opposed to males, this despite concessional rates provided by many government schemes. Planning sensitisation campaigns for greater awareness and disbursal schemes through friendly channels or platforms is imperative. While gender neutrality may seem far-fetched now, it is the only long-term solution for equal opportunities.

8.2.3 Necessity Versus Opportunity

Most women adopt entrepreneurial ventures due to a lack of opportunities in the formal workforce and not because they can spot opportunities in the marketplace. Traditionally, women face a lack of market access and networking opportunities as opposed to males, resulting in them turning to entrepreneurship as a necessity and not as an opportunity. The African Continental Free Trade Area agreement (AfCFTA) promises to provide women engaged in informal cross-border commerce (WICBT) with enhanced opportunities owing to the tariff reductions stipulated in the Protocol on Commerce in Goods. Women in agricultural value chains will get a comparative advantage by utilising the protocol for Rules of Origin, which allows access to inexpensive raw materials. It is aimed that women would achieve greater revenue increases and be better equipped to engage more significantly in AfCFTA intra-regional trade opportunities.

8.2.4 Male-Dominated Sectors (MDS) and Female Concentrated Sectors (FCS)

Female-owned businesses predominantly focus on commerce and retail sectors, whereas male-owned organisations prevail in manufacturing and agriculture. In manufacturing, areas predominantly occupied by males include metalworking, engineering, building, various manufacturing, and repair, as well as leather manufacturing and shoemaking/repair, except in Indonesia and Vietnam, where several of these sectors are predominantly female-orientated. It is a matter of observation that female-concentrated sectors (FCS) are less profitable compared to MDS. It is a

TABLE 8.1
Literature Analysis

Top Journals	No of Publications
Academy of Management Review	144
Entrepreneurship Theory and Practice	325
Journal of Business Venturing	290
American Journal of Sociology	51
Contemporary Sociology: A Journal of Reviews	73
SSRN Electronic Journal	545
Choice Reviews Online	146
Small Business Economics	310
Administrative Science Quarterly	70
Journal of Small Business Management	240

Source: Authors' compilations based on citations

matter of grave concern, as less profitability will make these sectors less attractive to females. It is further noted that women who cross over to the MDS show higher profitability in their enterprises. So if one aims to create a hierarchy of profits, it is generally the MDS that comes at number one, and women crossing over the MDS comes second, and the last is FCS. So the challenges are multifold, and policy directives are essential for the corrective actions. Over the years, a spurt in the academic literature related to entrepreneurship has been witnessed. Table 8.1 lists a few of the top journals publishing in the area with specific reference to women's entrepreneurship in developing countries.

8.3 RESEARCH METHODOLOGY

The Global Entrepreneurship Monitor (GEM) consortium publishes its annual data on various entrepreneurship-related parameters, both individual and specific to ecosystems in nations. This database holds statistical data for multiple world countries. This collection of information is an important framework to assess the attractiveness of entrepreneurship in various nations, also simultaneously comparing the perception related to the availability of the resources across the nations. Africa hosts a multitude of nations; however, due to the lack of participation of many countries, the data sets are either incomplete or not updated to the current years. Still, the data sets are relevant and important. The present study uses the GEM data set to analyse the following: Adult Population Survey (APS) is an extensive questionnaire administered to at least 2,000 adults in each GEM country, aimed at gathering detailed information regarding the entrepreneurial activities, attitudes, and goals of respondents.

Female/Male TEA Ratio: The proportion of the female population divided by the corresponding proportion for males aged 18–64 who are either early entrepreneurs or owner-managers of a 'new entrepreneurship.'

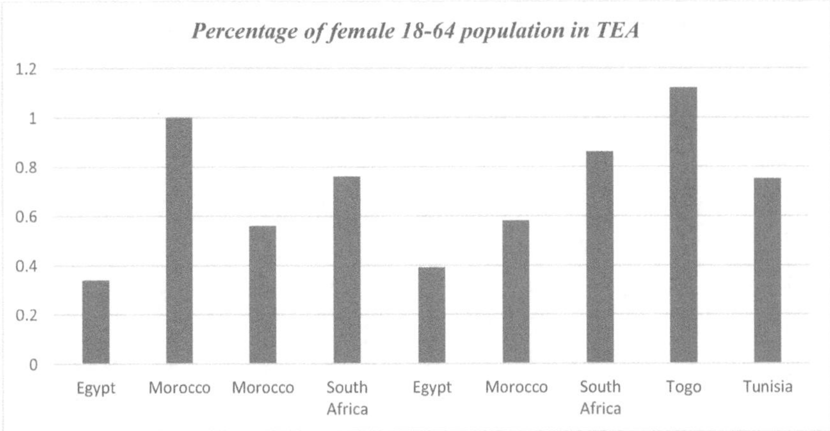

FIGURE 8.1 TEA ratio.

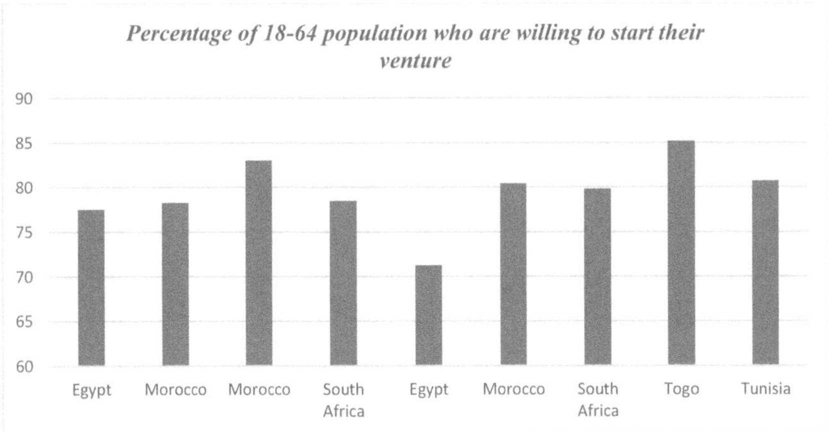

FIGURE 8.2 Intentions for entrepreneurship.

Entrepreneurship as an Intentional Career Choice Rate: The proportion of the 18–64 demographic that concurs with the assertion that, in their nation, the majority regard entrepreneurship as a favourable/intentional career option.

Governmental Policies—Support and Relevance: The extent to which the respective countries' public policies support entrepreneurship as a pertinent economic issue.

Entrepreneurial Education at Post School Stage: The extent to which training in creating or managing SMEs is incorporated within the education and training system in higher education such as vocational, college, business schools, and so on.

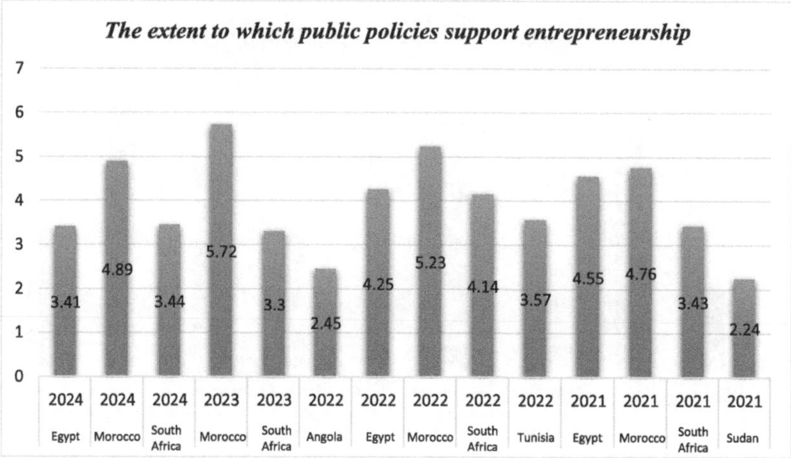

FIGURE 8.3 Government policies.

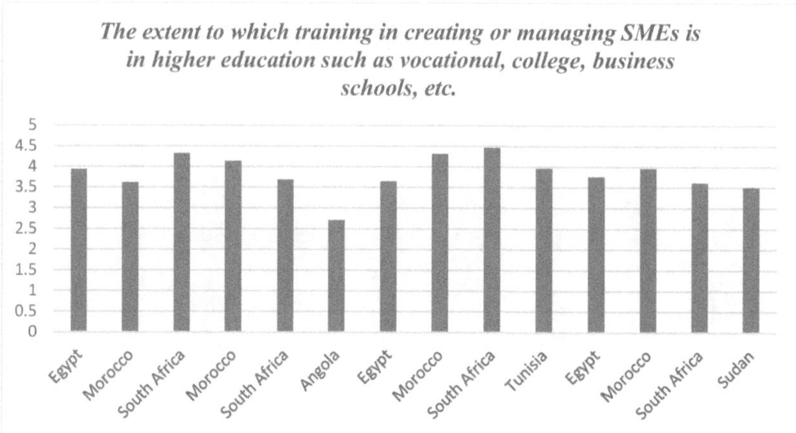

FIGURE 8.4 Training for SMEs.

8.4 DISCUSSION

African countries, despite being less developed, still show a significant amount of entrepreneurial inclinations, and well-documented reports from the World Bank, International Financial Corporations (IFC), and the International Monetary Fund (IMF) have shown encouraging statistics on the rate of women's entrepreneurship. A comparison with developed nations also shows higher rates of African women entrepreneurs. An analysis of various African countries on their 'intention to start a venture' shows a higher percentage of the population positive on their intention for enterprise creation. The GEM database is a compilation of the research statistics using an adult population survey (APS), i.e., using a sample of 18–64 from a country. Not all nations have participated in the GEM study based on the available

TABLE 8.2
Entrepreneurship Support Framework

Countries	Year	The extent to which public policies support entrepreneurship—entrepreneurship as a relevant economic issue (%)
Egypt	2024	3.41
Morocco	2024	4.89
South Africa	2024	3.44
Morocco	2023	5.72
South Africa	2023	3.3
Angola	2022	2.45
Egypt	2022	4.25
Morocco	2022	5.23
South Africa	2022	4.14
Tunisia	2022	3.57
Egypt	2021	4.55
Morocco	2021	4.76
South Africa	2021	3.43
Sudan	2021	2.24

Source: Authors' compilation based on GEM API.

databases on the GEM website. Two nations, particularly Morocco and Togo, have shown higher inclinations as compared to other nations. Total economic activity of women, which is calculated on the basis of the female over male ratio in terms of nascent enterprise creation. In this statistic as well, Morocco and Togo show nearly equal participation, which will result in a higher number of women entrepreneurs. It is interesting to note that both these countries are less developed as compared to other prominent nations like South Africa. It is imperative to study specific national problems to understand the drivers and challenges for women entrepreneurs. It is also evident a large number of women engage in the informal sector and are necessity-driven entrepreneurs. On the aspect of whether governments in respective nations understand entrepreneurship as a policy issue and provide support, most nations average below 5%, and only Morocco shows a higher percentage. As per the training for entrepreneurship or small enterprises in higher education institutes, most respondents have shown fewer initiatives are taken.

8.5 LIMITATIONS

The study is based on secondary and documented sources. While the sources used in the study are well respected and authentic, they portray an aggregative picture, and an analysis of country-specific issues is not discussed in detail. To understand the pertinent issues, it is important to uncover the ground-level issues, which is possible using primary database research. The limitation of the study is in its lack of exhaustive data to provide implications for the policymakers.

REFERENCES

Ojong, N., Simba, A., & Dana, L. (2021). Female entrepreneurship in Africa: A review, trends, and future research directions. *Journal of Business Research*, *132*, 233–248. https://doi.org/10.1016/j.jbusres.2021.04.032

Panda, S. (2018). Constraints faced by women entrepreneurs in developing countries: Review and ranking. *Gender in Management: An International Journal*, *33*(4), 315–331.

Skrbková, D., & Rydvalová, P. (2023). Entrepreneurial intentions: A comparative study of African and European university economics students. *Journal of African Business*, *25*(4), 792–810. https://doi.org/10.1080/15228916.2023.2276541

9 Diversity and the Role of Women Entrepreneurs in Africa *Challenges, Opportunities, and Social Impact*

Pallavi Padhy and Madhusmita Dash

9.1 INTRODUCTION

In today's era one of the most important critical drivers of economic growth and social transformation is women's entrepreneurship. It has been seen that women's entrepreneurship significantly contributed in a way that enhances gender equality as well as nurture communal improvement. In recent studies in African regions, it found that there is a great impact on women entrepreneur due to differences in social, cultural, and geographical factors. Across the African countries, one of the most important aspects is its diversity which influences the journey of women entrepreneurs. There are many factors that act as opportunities as well as challenges that women face in the journey of entrepreneurship like their qualification, the region they belong to, the sociocultural community they have, and the financial accessibility they can take. The development of these business run by women can reduce the social economic problems like poverty, illiteracy, and support equality.

Women entrepreneurs in Africa also face many barriers regarding economic and sociocultural challenges despite of their contributions. Barriers like lack of education, limited financial access, restricted training programmes, and unfair laws may affect the participation and involvement of women in the world of business. Supporting the women entrepreneurs not only boosts their personal growth but also improves the socio-economic well-being of the community of the African countries. This also leads to advancement of healthcare, social unity, and legal laws and policies.

The object of this research is how women's entrepreneurship is affected by diversity in African countries in terms of ethnical differences, socio-economical norms, and landscape of entrepreneurship. This also leads to analyse the role of women in advancing the gender quality and their social impact on improvement of community. With the help of existing literature, it focuses to analyse the key challenges and opportunities faced by women in running their businesses and provides recommendations for empowering women's entrepreneurship. Eventually, this paper shows if

DOI: 10.1201/9781003508953-9

there more involvement of women in the world of business will lift the economic conditions and provoke competitive advantages of the society in African countries.

9.2 LITERATURE REVIEW

According to this study, there are some elements that impact women's entrepreneurship in Africa which focus on not only the difficulties they face but also the opportunities for innovation and creations in running and expanding the business. Previous studies have shown how the diversity in Africa affects the women entrepreneur in various case studies. The following are some factors that are gathered from literature review to address the complications of women's entrepreneurship.

Studies have highlighted that *diversity*—in terms of socio-economic status, ethnicity, education, access to capital, and geography—is the focus for African women. Research by Amine and Staub (2009) highlights how African women's diverse cultural and social environments affect their access to resources and opportunities. Similarly, studies by Mordi et al. (2010) discuss the intersectionality of ethnicity and gender in shaping entrepreneurial outcomes for African women.

9.3 ETHNIC AND SOCIO-ECONOMIC FACTOR

Socio-economic status impacts African women's entrepreneurial opportunities by influencing their access to education and capital. Lower-income women often face more significant barriers due to a lack of financial resources or societal support (Gichuki et al., 2014). Ethnic affiliations can also either hinder or promote entrepreneurial ventures based on cultural norms and access to community networks (Robson et al., 2009). Women from minority ethnic groups or marginalized regions often face even greater hurdles in accessing opportunities.

Research highlights that socio-economic status and ethnic background significantly affect women's access to resources and networks essential for business success. For instance, women in many cases face problems related to education and limited access to capital for their growth in business (Kivung & Sweeney, 2023). Ethnic diversity within communities can either facilitate or obstruct entrepreneurship depending on the prevailing cultural attitudes towards women in business (Kara & Goktepe, 2023).

9.4 EDUCATION AND SKILL DEVELOPMENT

One of the main factors for women's entrepreneurship is education. According to studies, women with higher education levels tend to have more business success due to greater access to information, networks, and skills (Nichter & Goldmark, 2009). However, many African women, particularly in rural areas, experience limited access to education and training, which can reduce their competitiveness in the market. Education emerges as a pivotal determinant of entrepreneurial success. Programmes aimed at improving educational access for women have shown promise in enhancing their entrepreneurial skills and business acumen. A study by the International

Finance Corporation (2023) notes that women who receive entrepreneurial training exhibit higher business performance and innovation levels (International Finance Corporation, 2023).

9.5 ACCESS TO FINANCIAL CAPITAL

Research by Fletschner and Kenney (2011) emphasizes that financial institutions often neglect women entrepreneurs, limiting their ability to scale businesses. Women entrepreneurs face challenges related to creditworthiness, which is influenced by discriminatory practices and limited collateral.

The lack of access to financial resources is a well-documented barrier. Women entrepreneurs often encounter difficulties in securing funding due to systemic biases in lending practices. However, initiatives like gender-lens investing and targeted microfinance programmes are emerging to address these disparities, thereby enabling women to access the necessary capital for their businesses (She WINS Africa, 2023).

9.6 BROADER SOCIAL IMPACT

Numerous studies have documented the social benefits of women's entrepreneurship in Africa. Women entrepreneurs play a key role in addressing poverty, as they often reinvest profits back into their families and communities (Mayoux, 2001). Community development is another area where women entrepreneurs contribute significantly by improving local infrastructure and supporting education and healthcare initiatives (Charman et al., 2012).

Research also demonstrates how women entrepreneurs advance gender equality by challenging traditional gender roles and becoming role models in their communities (Stevenson & St-Onge, 2005). Women entrepreneurs are increasingly recognized for their role in driving social change. Their contributions extend beyond economic activities to include improvements in community development, education, and healthcare. Studies indicate that women-led businesses often reinvest profits back into their communities, which is critical for poverty alleviation and advancing gender equality (UNESCO, 2023).

9.7 POLICY AND PRIVATE SECTOR INITIATIVES

Existing literature highlights that targeted policy interventions and private sector initiatives can play a key role in reducing barriers for African women entrepreneurs. Studies emphasize the importance of providing access to education, capital, and networks to empower women (Halkias et al., 2011). Policies that address gender-biased financial systems and legal frameworks are also crucial, as they can create more equitable opportunities for women to participate in the entrepreneurial ecosystem.

Recent literature emphasizes the need for supportive policy frameworks and private sector initiatives that empower women entrepreneurs. Recommendations include enhancing access to education, capital, and networks, as well as implementing legal reforms to remove barriers faced by women (Kara & Goktepe, 2023).

9.8 GEOGRAPHICAL DISPARITIES

Geographic factors, particularly for women in rural areas, create additional barriers to entrepreneurship. Studies (World Bank, 2020) show that urban women tend to have better access to markets, infrastructure, and financial institutions, while rural women face isolation, lack of market access, and inadequate support systems.

The literature review reveals a dual reality for African women entrepreneurs: significant barriers exist due to socio-economic and cultural challenges, but there are also unique opportunities for them to contribute to both *economic growth* and *social transformation*. Addressing these barriers through education, capital access, and inclusive policies is essential to fostering a more sustainable and inclusive economic landscape across Africa.

In summary, while African women entrepreneurs face significant barriers due to socio-economic, educational, and cultural factors, they also have unique opportunities to drive social and economic change. An inclusive entrepreneurial ecosystem that addresses these challenges can facilitate women's contributions to economic growth and social transformation across the continent.

9.9 ADDRESSING THE GAP

Despite the increasing recognition of the importance of women entrepreneurs in Africa, several research gaps remain that warrant further exploration.

Intersectionality of Challenges: While existing studies address various barriers faced by women entrepreneurs, there is limited research on the intersectionality of these challenges. Factors such as ethnicity, socio-economic status, and geographical location often interact in complex ways, influencing the entrepreneurial experiences of women. Further investigation into how these intersections affect access to resources and opportunities could lead to more ways to overcome the challenges faced by diverse groups of women (Kivung & Sweeney, 2023).

Impact of Policy Frameworks: Although several studies advocate for policy interventions to support women entrepreneurs, there is a lack of empirical research evaluating the effectiveness of existing policies across different African countries. Understanding how specific policy measures impact women's entrepreneurship would help identify best practices and areas needing improvement (Kara & Goktepe, 2023).

Longitudinal Studies: Much of the existing literature is cross-sectional, capturing a snapshot of women's entrepreneurship at a particular time. Longitudinal studies that track women entrepreneurs over time could shed light on how their businesses evolve, how their challenges change, and what factors contribute to long-term success (International Finance Corporation, 2023).

Role of Technology and Innovation: While technology is recognized as a tool for empowerment, more research is needed on how women entrepreneurs leverage technology in their businesses and the barriers they face in accessing digital resources. Understanding the role of innovation and technology in enhancing women's entrepreneurship could reveal new avenues for support and development (UNESCO, 2023).

Cultural Norms and Perceptions: The influence of cultural norms on women's entrepreneurship is under-researched. Exploring how societal attitudes towards women in business impact entrepreneurial intentions and behaviours could provide insights into the social dynamics at play (She WINS Africa, 2023).

Measurement of Social Impact: Although the social impact of women entrepreneurs is acknowledged, there is a lack of systematic frameworks to measure this impact. Research that develops metrics for assessing the contributions of women-owned businesses to community development, poverty alleviation, and gender equality would enhance the understanding of their broader societal role (Kara & Goktepe, 2023).

Identifying these gaps can lead to consider the landscape of women's entrepreneurship in Africa, enabling stakeholders to implement effective interventions that foster an inclusive and supportive environment for female entrepreneurs.

9.10 RECOMMENDATIONS

Based on the identified research gaps and the existing literature on women's entrepreneurship in Africa, the following recommendations can be made to enhance support for female entrepreneurs and foster a more inclusive entrepreneurial ecosystem.

Develop Intersectional Policies: Policymakers should create and implement intersectional policies that recognize the diverse challenges faced by women entrepreneurs based on their socio-economic status, ethnicity, and location. Some policies and programmes need to be taken care of to address the need of the women entrepreneurs.

Evaluate Existing Policies: There is a need for rigorous evaluation of existing policies aimed at supporting women entrepreneurs. Governments and organizations should conduct impact assessments to understand the effectiveness of these policies and identify areas for improvement. This evidence-based approach can inform future policy development and implementation (Kara & Goktepe, 2023).

Promote Digital Literacy and Access: As technology becomes increasingly important in entrepreneurship, initiatives aimed at improving digital literacy among women should be prioritized. Training programmes that focus on technology use in business, combined with efforts to enhance access to digital tools and resources, can empower women entrepreneurs and enable them to compete effectively in the market (International Finance Corporation, 2023).

Foster Networking and Mentorship: Establishing networks and mentorship programmes can facilitate knowledge sharing and resource access for women entrepreneurs. Programmes that connect experienced entrepreneurs with newcomers can provide valuable guidance and support, helping to navigate the challenges of starting and growing a business (She WINS Africa, 2023).

Measure Social Impact Systematically: Developing standardized metrics to assess the social impact of women-owned businesses can provide valuable insights into their contributions to community development and poverty alleviation. This data can inform policy decisions and highlight the importance of investing in women's entrepreneurship (Kara & Goktepe, 2023).

Encourage Private Sector Involvement: The private sector should play a more active role in supporting women entrepreneurs through targeted investments and initiatives. Gender-lens investing, which prioritizes funding for women-led businesses, can help address the funding gap and promote gender equity in entrepreneurship (UNESCO, 2023).

9.11 DISCUSSION

The discussion around women's entrepreneurship in Africa highlights the significant barriers that female entrepreneurs face, as well as the potential they hold for driving economic and social change. While progress has been made in recognizing the contributions of women entrepreneurs, challenges such as access to funding, education, and networks continue to persist. Addressing these issues requires a multifaceted approach involving government, private sector, and community initiatives.

The emphasis on intersectionality is crucial, as the experiences of women entrepreneurs are not monolithic. Factors such as ethnic background, geographic location, and socio-economic status create diverse entrepreneurial landscapes that must be understood and addressed. This recognition can lead to more effective policy measures that cater to the unique needs of various groups, ultimately fostering a more inclusive environment for all women entrepreneurs.

Moreover, the role of technology and innovation cannot be overstated. As the digital economy continues to expand, equipping women with the necessary skills and resources to leverage technology in their businesses will be vital for their success and sustainability. This transformation can also play a key role in addressing broader social issues, such as gender inequality and economic disparity.

In conclusion, fostering an inclusive entrepreneurial ecosystem in Africa requires concerted efforts across various sectors. By implementing targeted policies, enhancing access to resources, and promoting collaboration, African countries can empower women entrepreneurs to contribute significantly to economic growth and social transformation.

9.12 CONCLUSION

In conclusion, diversity plays a critical role in shaping women's entrepreneurship in Africa, presenting both challenges and opportunities. This review has highlighted the multifaceted barriers that female entrepreneurs encounter, including socio-economic disparities, cultural norms, and limited access to education and financial resources. However, it has also emphasized the potential for women to drive significant social and economic change within their communities through their entrepreneurial endeavours.

Addressing the identified research gaps—such as the need for intersectional policy frameworks, evaluations of existing initiatives, and systematic measurements of social impact—is essential for creating a more inclusive entrepreneurial ecosystem. By fostering environments that support education, access to capital, and technology use, African countries can empower women entrepreneurs, enabling them to contribute to economic growth and promote social transformation effectively.

Investing in women's entrepreneurship not only enhances economic development but also fosters gender equality and community well-being. Ultimately, a concerted effort from governments, private sectors, and civil society is necessary to unlock the full potential of women entrepreneurs in Africa, creating sustainable and inclusive growth for the continent's future. The continued support and empowerment of women in business will lead to significant benefits for families, communities, and the broader economy.

REFERENCES

Amine, L. S., & Staub, K. M. (2009). Women entrepreneurs in sub-Saharan Africa: An institutional theory analysis from a social marketing point of view. *Entrepreneurship & Regional Development*, *21*(2), 183–211. https://doi.org/10.1080/08985620802182144

Charman, A. J., Petersen, L. M., & Piper, L. (2012). Informal small businesses in townships: A case study of the role of business owners' ethnic identity and cultural capital in shaping business practices. *Journal of Developmental Entrepreneurship*, *17*(1), 1250001. https://doi.org/10.1142/S108494671250001X

Fletschner, D., & Kenney, L. (2011). Rural women's access to financial services: Credit, savings, and insurance. *Agriculture and Rural Development*, *58*(5), 67–71.

Gichuki, C. N., Njeru, A., & Tirimba, O. I. (2014). Challenges facing micro and small enterprises in accessing credit facilities in Kangemi harambee market in Nairobi city county, Kenya. *International Journal of Scientific and Research Publications*, *4*(12), 1–25.

Halkias, D., Nwajiuba, C., Harkiolakis, N., & Caracatsanis, S. M. (2011). Challenges facing women entrepreneurs in Nigeria. *Management Research Review*, *34*(2), 221–235. https://doi.org/10.1108/01409171111102821

International Finance Corporation. (2023). *She WINS Africa: Unlocking the potential of women-owned startups in Africa*. Retrieved from https://www.ifc.org/

Kara, A., & Goktepe, H. (2023). Ethnic diversity and women entrepreneurship in Africa: Impacts and opportunities. *Journal of African Business*, *24*(3), 325–343. https://doi.org/10.1080/15228916.2023.2156897

Kivung, D., & Sweeney, K. (2023). Barriers and opportunities for women entrepreneurs in Africa: A socio-economic perspective. *African Journal of Business Management*, *17*(4), 512–530. https://doi.org/10.5897/AJBM2023.12345

Mayoux, L. (2001). Tackling the down side: Social capital, women's empowerment and micro-finance in Cameroon. *Development and Change*, *32*(3), 435–464. https://doi.org/10.1111/1467-7660.00212

Mordi, C., Simpson, R., Singh, S., & Okafor, C. (2010). The role of cultural values in understanding the challenges faced by female entrepreneurs in Nigeria. *Gender in Management: An International Journal*, *25*(1), 5–21. https://doi.org/10.1108/17542411011019904

Nichter, S., & Goldmark, L. (2009). Small firm growth in developing countries. *World Development*, *37*(9), 1453–1464. https://doi.org/10.1016/j.worlddev.2009.01.013

Robson, P. J. A., Haugh, H. M., & Obeng, B. A. (2009). Entrepreneurship and innovation in Ghana: Enterprising Africa. *Small Business Economics*, *32*(3), 331-350. https://doi.org/10.1007/s11187-008-9121-2

She WINS Africa. (2023). Women's innovation and entrepreneurship for social impact in sub-Saharan Africa. *She WINS Africa*. Retrieved from https://www.shewinsafrica.org/

Stevenson, L., & St-Onge, A. (2005). *Support for growth-oriented women entrepreneurs in Ethiopia, Kenya and Tanzania: An overview report*. International Labour Organization.

UNESCO. (2023). *Almost one in four female African entrepreneurs have innovated*. Retrieved from https://www.unesco.org/

10 Understanding the Factors Influencing Women's Entrepreneurship in Maharashtra

A Comprehensive Analysis

Hrishikesh Kakde and
Pravinkumar Govind Shastri

10.1 INTRODUCTION

A key factor in the prosperity and well-being of society is entrepreneurship. It is also a powerful force for economic expansion. It fosters the vital innovation needed to address some of society's biggest issues, such the sustainable development goals (SDGs) of the UN or shocks from various international events, in addition to embracing new possibilities, boosting productivity, and creating jobs [1].

Gender equality, poverty reduction, and sustainable economic development all depend on women's involvement in the economy. Women entrepreneurs also continue to be key drivers of economic growth and social development in their nations and local communities. They bring fresh perspectives, concepts, and inventions to the markets they serve. In India, women make up 17% of the GDP, which is less than half of the average for the world [2].

Also, it appears that women's entrepreneurship has an impact on social and environmental issues in addition to economic progress. Women-led enterprises frequently place a high priority on environmental sustainability, open reporting, and community well-being, using creative solutions to solve significant issues [3]. Given that women work extensively not only to maintain their families but also to ensure their financial well-being, numerous scholars consider women's entrepreneurship to be a 21st-century phenomenon [4]. Opinions on women's empowerment through an array of activities vary, but in ancient India, women enjoyed equal status with men. Later, however, a number of social and economic factors caused women to become increasingly disadvantaged. The Indian constitution, which had been adopted at the

DOI: 10.1201/9781003508953-10

time, cleared up an abundance of uncertainty regarding women's status in India, which helped women gain ground there.

According to the researcher, women's entrepreneurship represents more than simply a way for women making money; it also signifies how women are treated in society and how they gain acceptance in roles that are equal to those of men. Over the past few decades, women's entrepreneurship has gained significant attention in study, making it a rising global phenomenon [5]. It is acknowledged as a source of growing entrepreneurial diversity in a variety of economic scenarios in addition to helping economies in terms of job creation and economic growth [6],[7]. Globally, women entrepreneur is not anything new; one in six of them stated that they planned to start an enterprise soon. About 28% of women indicated that they intended for launching a business, which was the highest rate of entrepreneurial intention seen in low-income nations [1]. Nonetheless, the issue with India is that, according to the 2014 Sixth Economic Census, just 13% of Indian women are employed, and only 14% of them own or manage enterprises. Over 90% of women-owned businesses are microenterprises, with approximately 79% being self-financed. In India, women make up barely 17% of the GDP, which is less than half of the global average [8]. Indian women entrepreneurs experience discrimination despite the entrepreneurial rosy picture, and their earnings hardly cover subsistence. To assist them in identifying the most relevant components of their performance and motivating elements, as well as motivation, scholars have examined their entrepreneurship traits, such as business competence, performance, and entrepreneurial orientation. To increase knowledge of the initiatives that need to be implemented to overcome their entrepreneurial obstacles, other research have concentrated on mapping and classifying implications pertaining to their managerial, practical, and policy elements [9]. There are several characteristics that contribute to women being entrepreneurs; these include being wealthy, being drawn to the field by traits like education and a willingness to take risks, facing financial difficulties, being an entrepreneur in a remote area, and being highly successful. Numerous limitations are applied when examining women's entrepreneurship. For instance, certain business segments—such as subcontractorships and service-oriented enterprises—are recognized as areas of women's entrepreneurship. The main cause, according to researchers, is that many Indians still do not understand what formal and professional entrepreneurship means. Both the scope of control and the division of labor remain on paper. The historical status of women has left differing starting conditions for the sexes, which contributes to the gender imbalance in business development around the world [10]. In addition, entrepreneurship is influenced by a variety of human elements, including psychological traits, a desire for personal development, risk tolerance, self-assurance, and the exploration of potential entrepreneurial endeavors [11].

Certain studies also take into account socioeconomic criteria such as age, education, income, and professional experience as individual factors impacting female entrepreneurship. In addition to these, societal and institutional elements that support female entrepreneurship can be identified through the theoretical review. They connected with the surroundings that women live in. Studies highlight the importance of families in particular [10]. Women's levels of self-confidence are influenced by the relationships they have with organizations, business networks, and support agency through their employment [12]. The view of one's abilities, social networks, and family role are among the "informal ways" that influence women's entrepreneurship [13]. Research has indicated the impact that the current economic environment has on women's entrepreneurship from the start of the crisis in the early twenty-first century. Based on this, the economic and employment aspects that account for female entrepreneurship might be ranked third. They have to do with the dearth of work prospects, high unemployment rates, labor market volatility and uncertainty, and discrimination against women in the workplace [10]. According to data from the Observatory Global Entrepreneurship Monitor (GEM), women have become more entrepreneurial in recent years as a means of securing personal income, with this increase being particularly notable among this group [14]. The education and training of women as entrepreneurs is one of the major influencers at the outset of entrepreneurship. Women are very involved in household chores, and research indicates that when household responsibilities decrease, entrepreneurship flourishes. One of the factors that determines the beginning of a firm is an external financial resource; they can act as either facilitators or obstacles for the growth of female entrepreneurship. It's widely acknowledged that women typically face greater challenges in accessing funding opportunities. Various factors contribute to this phenomenon. For instance, women-owned businesses often operate on a smaller scale, requiring less capital. Additionally, women tend to have lower levels of ambition for growth compared to their male counterparts [15].

10.2 METHODOLOGY

This study wanted to explore more on women in business, specifically in Maharashtra. It aimed to gather information about what factors are responsible for women being not becoming entrepreneur or choosing entrepreneurship. The researchers hoped their work would encourage others to explore these ideas further.

To get this information, the researchers observed postgraduate female students and surveyed female respondents of labor force in the region through convenient sampling. The sample of 100 could be one of the limitations of the study; yet considering the time constraints the researcher analyzed a sample of only100. They sent out surveys using email and a Google Form. Even though they sent 320 surveys, only 100 women responded.

The aim was to find out the entrepreneurial intention and factors affecting entrepreneurial intention of women.

10.3 RESULTS AND DISCUSSIONS

10.3.1 Demographic Profile of the Respondents

From the data collected 18% were of age less than 20, 10% were aged between 21 years and 30 years, 45% were in the range of 31–40 years, 20% were of 41–50 years age, and 7% were of more than 50 years of age (Table 10.1). The first two majorities were college-going students. Forty-eight percent respondents belong to rural

TABLE 10.1
Demographic Profile of the Respondents (n=100)

Variables		N	%
Age group	<20	18	18%
	21–30	10	10%
	31–40	45	45%
	41–50	20	20%
	>50	07	07%
Place of origin	Rural	48	48%
	Semi-urban	12	12%
	Urban	40	40%
Marital status	Single	21	21%
	Married	79	79%
Business history	Yes	35	35%
	No	65	65%
Experience	<1 year	16	16%
	1–5 years	68	68%
	5–10 years	11	11%
	10 years	05	05%
Enterprise industry	Retail shop	12	12%
	Fabric and garments	10	10%
	Food and beverages	08	08%
	Beauty shops	25	25%
	Jewelry	05	05%
	Pharmacy	08	08%
	Other	32	32%

Source: (Primary data)

area, 12% to semi-urban area, and 40% belongs to urban area. Twenty-one percent respondents were single, and 79% were married. Thirty-five percent respondents had business history in various areas and 65% had no business history. Sixteen percent respondents had less than one year experience, 68% had 1–5 years' experience in handling entrepreneurial activities, 11% with 5–10 years of experience, and 5 percent with over 10 years' experience. Twelve percent of the respondents dealt in retail shops, followed by beauty shops with 25% respondents, 10% were in fabric and garments industry, 8% in food and beverages and pharmacy respectively, 5% in jewelry, and 32% in other forms of businesses. The 'other' criteria in this case are largely dominated by agricultural-related activities correlating with the rural area.

10.3.2 Adoption of Digitalization

The researcher surveyed the respondents, 52% of the existing businesses have adopted the digital payment methods, 35% respondents heavily rely on digital marketing, 8%

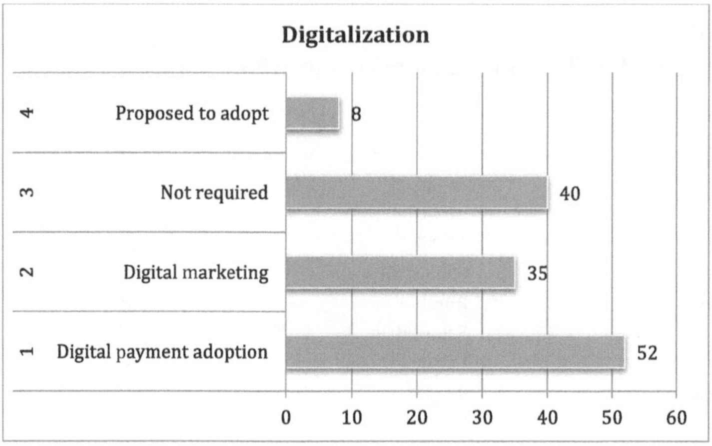

FIGURE 10.1 Adoption of digitalization.

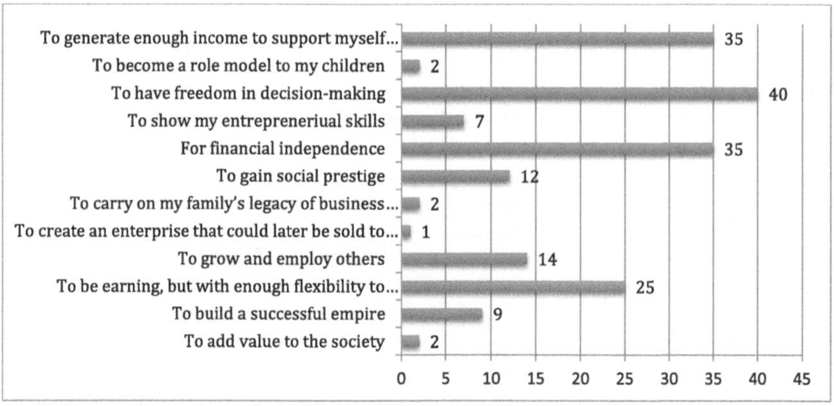

FIGURE 10.2 Motivation for entrepreneurship.

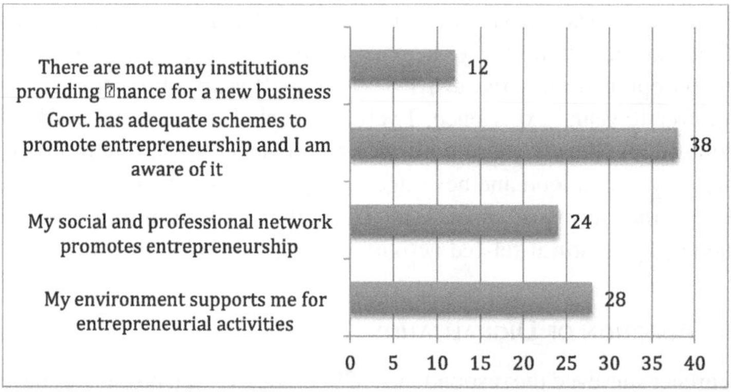

FIGURE 10.3 Socioeconomic factors influencing entrepreneurial intention.

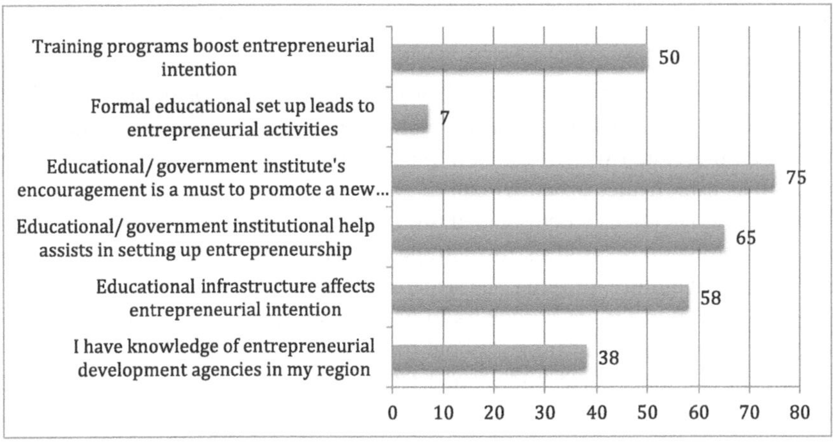

FIGURE 10.4 Educational and institutional support influencing entrepreneurial intention.

TABLE 10.2
Factor Analysis Table

Rotated Component Matrix[a]

	Component		
	1	2	3
Education		.899	
Income		.845	
Professional experience		.835	
Desire for growth	.860		
Risk level	.859		
Psychological aspects			.846
Freedom in decision-making			.781
Financial independence			.825
Self-confidence	.852		

Extraction method: principal component analysis.
Rotation method: varimax with Kaiser normalization.

[a] Descriptive Statistics

said they are planning on adoption of digital tools in different areas, and 40% said it is not required.

10.3.3 Motivation to Be an Entrepreneur

The adopted questionnaire also asked about the motivation to be an entrepreneur, it was found that freedom in decision-making was the most dominant response followed by income generation ability and financial independence.

10.3.4 Socioeconomic Factors Influencing Entrepreneurial Intention

The study continued to dig into the contextually related factors to find out their impact on entrepreneurial intention. It was found that relational support is key to building entrepreneurial intention wherein relational support means social connections of an individual which results in positive in terms of creating entrepreneurial intention as it motivates an individual to go for it. 'The one who is thirsty will find the well' exactly applies when sociocultural factors are seen, as the structural factors do not majorly help in building the entrepreneurial intention but assist in promoting entrepreneurial culture. In India, the structural changes are happening rapidly which is a very good sign for entrepreneurs to be. The study depicts that there should be more focus on creating awareness about these structural factors in association with educational institutes, local bodies, and NGOs working for it which might lead to motivate the entrepreneurs to be.

10.3.5 Educational and Institutional Support Influencing Entrepreneurial Intention

When questions related to educational and institutional supports were asked, 75% of the respondents were of the opinion that educational and/or government encouragement is must when it comes to first-generation entrepreneur, followed by assistance from them with support during educational stage is crucial for generating entrepreneurial intention. Yet there was a lack of information on agencies promoting entrepreneurship in the region.

Along with the important variables mentioned earlier, detailed factor analyses were done using SPSS software, the result of which are shown in Table 10.2.

10.3.6 Rotation Converged in Four Iterations

The data collected for analysis holds all the assumptions of Exploratory Factor Analysis. In FACTOR 1, the variables 4, 7, and 8 have high loadings of 0.860, 0.781, and 0.825 on FACTOR 1. This indicates that FACTOR 1 is a combination of these three variables. Therefore, FACTOR 1 can be interpreted as a combination of 'Desire for growth', 'Freedom in decision-making' and 'financial independence'. Hence, FACTOR 1 can be named as 'NECESSITY'.

In the rotated factor Matrix, we noticed that variables 1, 2, and 3 have loadings of 0.899, 0.845, and 0.835 on FACTOR 2, respectively. This suggests that FACTOR 2 is a combination of these three variables. This FACTOR 2 can be interpreted as a combination of 'Education', 'Income', and 'Professional experience'. In this case, FACTOR 2 can be named as 'SOCIO-ECONOMIC'.

In FACTOR 3, the variables 5, 6, and 9 have high loadings of 0.859, 0.846, and 0.852 on FACTOR 3. This indicates that FACTOR 3 is a combination of these three variables. Therefore, FACTOR 3 can be interpreted as a combination of 'Risk Tolerance Level', 'Psychological Aspects', and 'Self-Confidence'. Hence, FACTOR 3 can be named as 'PERSOANL FACTORS'.

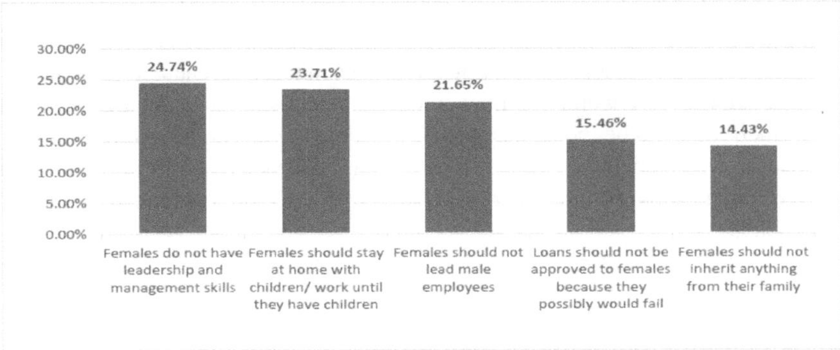

FIGURE 10.5 Most discriminating statements against women entrepreneurs.

(Source: Ahmetaj, Bardhyl, Alba Demneri Kruja, and Eglantina Hysa. 2023. Women Entrepreneurship: Challenges and Perspectives of an Emerging Economy. Administrative Sciences 13: 111. https://doi.org/10.3390/admsci13040111)

Despite of it, women entrepreneurs' face difficulties in different areas, some of them are as follows:

Literature, researcher's observations, and Figure 10.5 depict that women entrepreneurs are criticized on the basis of leadership and management skills, household activities as perceived priority, patriarchy, financial incredibility, and share in the family resources. This could be a leading research topic to dig into the details.

10.4 CONCLUSION

The factors explored in this study were hardly studied earlier; these factors can lead to extending support by government agencies leading to enhancing women's entrepreneurship in Indian context. The main aim of this research was to explore the factors impacting women's entrepreneurship; it was found during the study that necessity, socioeconomic factors, and personal factors have a significant impact on women's entrepreneurship. Further, the perceived approach is yet the major constraint in the empowerment of women through entrepreneurship. The 'other' sector mentioned in the research is mostly dominated by agricultural and related entrepreneurial activities, which restrict women entrepreneurs to local levels and up to substantial activities rather than commercialization to greater extent. Beauty shops, retails, and food and beverages are found to be the undertaken activities by the women entrepreneur respondents. Women entrepreneurs are leading through digitalization, but the scope is limited to only digital advertisement/promotion and mode of payment. The willingness to exhibit life through freedom is the biggest motivation for women's entrepreneurship. Financial independence, supporting family, work–life balance are found to be some other factors motivating women to take up entrepreneurial activities. Women entrepreneurs seek social and government support for encouraging and promoting their entrepreneurial activities. Women respondents are of strong opinion that educational and institutional support would facilitate the entrepreneurial activities. Training, formal education leading to entrepreneurship, and financial assistance are found crucial for empowering women's entrepreneurship. Since families still don't involve women in financial decisions, it shows acceptance at societal level, which stands to be the biggest hurdle in women's entrepreneurship. This factor also leads to women being self-exploring holding them back in exploring institutional and government support for entrepreneurial activities. Good governance may resolve this problem.

REFERENCES

[1] Global Entrepreneurship Monitor, "Global Entrepreneurship Monitor Research Highlights Significant Increase in Growth-oriented Women Entrepreneurs Worldwide," GEM Global Entrepreneurship Monitor, Nov. 16, 2021. https://www.gemconsortium.org/reports/womens-entrepreneurship

[2] M. Consulting, "Decoding Government Support to Women Entrepreneurs in India," 2022. Available: https://www.niti.gov.in/sites/default/files/2023-03/Decoding-Government-Support-to-Women-Entrepreneurs-in-India.pdf

[3] D. Khanna, "Women Entrepreneurs are Key to India's Green Economy," Forbes, Mar. 07, 2024. https://www.forbes.com/sites/deepalikhanna/2024/03/07/women-entrepreneurs-are-key-to-indias-green-economy-deepali-khanna/?sh=6961071c6793 (accessed Apr. 23, 2024).

[4] Rashid, S., and Ratten, V., "A Systematic Literature Review on Women Entrepreneurship in Emerging Economies While Reflecting Specifically on SAARC Countries," springerprofessional.de. https://www.springerprofessional.de/en/a-systematic-literature-review-on-women-entrepreneurship-in-emer/17546618 (accessed Apr. 23, 2024).

[5] Henry, C., Foss, L., and Ahl, H., "Gender and Entrepreneurship Research: A Review of Methodological Approaches," International Small Business Journal: Researching Entrepreneurship, vol. 34, no. 3, pp. 217–241, Jan. 2015, doi: https://doi.org/10.1177/0266242614549779.

[6] Kelley, D. J., Bosma, N., and Amoros, J. E., "GEM Global Entrepreneurship Monitor," GEM Global Entrepreneurship Monitor, 2010. https://www.gemconsortium.org/report/gem-2010-global-report

[7] Verheul, I., Stel, A. V., and Thurik, R., "Explaining Female and Male Entrepreneurship at the Country Level," Entrepreneurship & Regional Development, vol. 18, no. 2, pp. 151–183, Mar. 2006, doi: https://doi.org/10.1080/08985620500532053.

[8] World Economic Forum, "Global Gender Gap Report 2023," World Economic Forum, Jun. 2023. https://www3.weforum.org/docs/wef_gggr_2023.pdf

[9] Corrêa, V. S., Brito, F. R. da S., de Lima, R. M., and Queiroz, M. M., "Female Entrepreneurship in Emerging And Developing Countries: A Systematic Literature Review," International Journal of Gender and Entrepreneurship, Dec. 2021, doi: https://doi.org/10.1108/ijge-08-2021-0142.

[10] Pérez-Pérez, C., and Avilés-Hernández, M., "Explanatory Factors of Female Entrepreneurship and Limiting Elements," Suma de Negocios, vol. 7, no. 15, pp. 25–31, Jan. 2016, doi: https://doi.org/10.1016/j.sumneg.2015.12.004.

[11] Cromie, S., "Motivations of Aspiring Male and Female Entrepreneurs," Journal of Organizational Behavior, vol. 8, no. 3, pp. 251–261, Jul. 1987, doi: https://doi.org/10.1002/job.4030080306.

[12] Ventura Fernández, R., and Quero Gervilla, M. J., "Factores Explicativos de la intención de emprender en la mujer. Aspectos diferenciales en la población universitaria según la variable género," Cuadernos de Gestión, vol. 13, no. 1, pp. 127–149, 2013, doi: https://doi.org/10.5295/cdg.100271rv.

[13] Alvarez, C., Noguera, M., and Urbano, D., "Condicionantes Del Entorno Y Emprendimiento Femenino. Un Estudio Cuantitativo En España." Accessed: Apr. 24, 2024. [Online]. Available: https://www.mintur.gob.es/Publicaciones/Publicacionesperiodicas/EconomiaIndustrial/RevistaEconomiaIndustrial/383/Claudia%20Alvarez.pdf
[14] Nani, G. V., and Mathwasa, J., "Factors that Discourage Highly Educated Women from Being Entrepreneurs," European Journal of Management and Marketing Studies, vol. 8, no. 1, Jan. 2023, doi: https://doi.org/10.46827/ejmms.v8i1.1403.
[15] Marlow, S., "A Safety Net or Ties that Bind? Women, Welfare and Self-Employment," International Journal of Sociology and Social Policy, vol. 26, no. 9/10, pp. 397–410, Sep. 2006, doi: https://doi.org/10.1108/01443330610690541.

11 AI-Enhanced Women's Empowerment in Africa

Driving Sustainable Business Practices through Green HRM Technologies

Jan Alexa Sotto, Jonah Tomno, Sheela Bijlwan, Sonam Bhadauriya, and Vijay Prakash Srivastava

11.1 INTRODUCTION

According to projections, artificial intelligence (AI) will add US$15.7 trillion to the global GDP by 2030, of which US$6.6 trillion would come from higher productivity and US$9.1 trillion from consumption effects. With a predicted annual growth rate (CAGR 2024–2030) of 17.68%, the AI market in Africa is expected to reach US$6.90 billion in 2024 and US$18.33 billion in volume by 2030 [1].

The value of social, ethical, and environmental aims is being recognized by businesses more and more [1]. Organizations are putting more emphasis on human performance and development, as well as on the health of communities and the environment, than just financial benefit. One industry that supports "green" enterprises is sustainable human resource management (SHRM). The sustainable development goals (SDGs) are achieved through the organization's human capital adopting new ecological practices and using cutting-edge sustainable tactics [2]. AI is always working to meet the demands of Industry 4.0 by transforming traditional enterprises into "smart factories" where human labor may be used to achieve organizational sustainability with the least amount of human effort [3]. The concept of "Green HRM" has grown in acceptance throughout the world.

To increase productivity, cut expenses, and increase employee engagement and retention, GHRM takes part in eco-friendly projects and events [4]. Businesses are already encouraging green HRM practices, personal business accountability, and social awareness [5, 6,7]. The HRM functions assist sustainable development within the organization by using "Eco-focused" Green HRM practices and laws [8, 9,10]. All HRM processes, such as hiring and selection, training and assistance, pay and incentives, evaluation methods, and terms and conditions of employment, among others, are included in green HRM. Strategies and practices for sustainable green HRM

 DOI: 10.1201/9781003508953-11

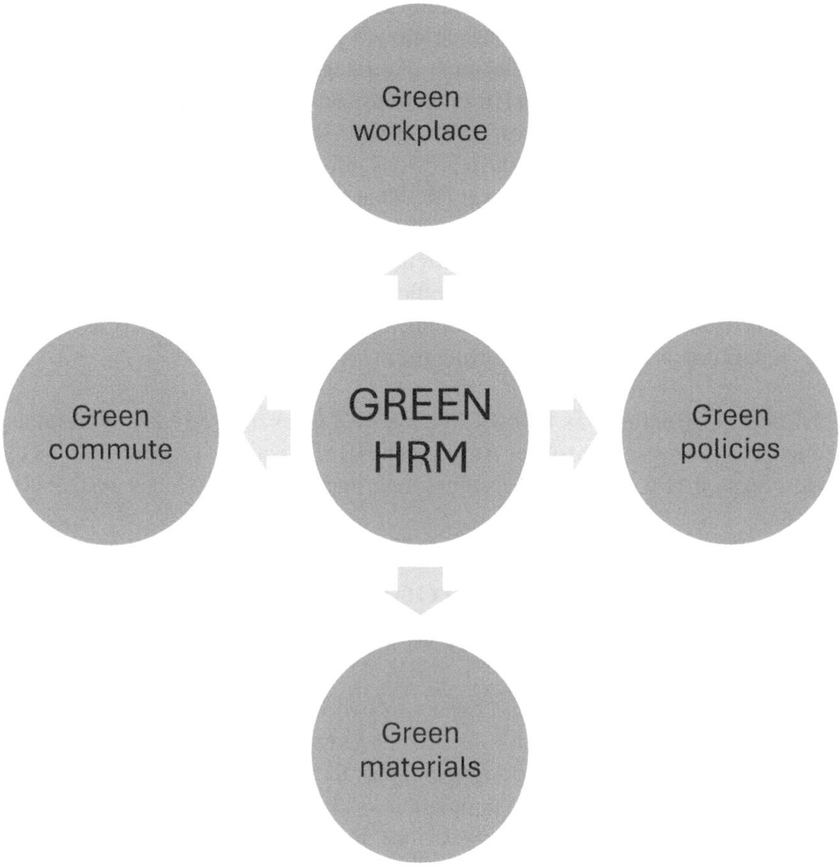

FIGURE 11.1 Green HRM.

encourage the long-term usage of wealth within the organization while enhancing employee pride [11].

AI supports businesses through the use of cutting-edge digital technologies, cloud computing, data storage, decision-making apps, and smart analytical tools. Use AI to craft sustainable business strategies and greener HRM across the organization, turning insights into measurable resource efficiency and lower emissions. Apply data-driven workforce planning to advance fair hiring, smart scheduling, learning pathways, and compliant, eco-conscious policies [12–15]. Worker job happiness, well-being, and interpersonal relationships all grow with improved AI and green HRM. Green HRM policies cover all of the following: pay and management incentives, job security, and collaboration to boost output. A study demonstrates some of the advantages of green HRM (Figure 11.1).

In this study the researcher emphasized on the application of AI to the adoption and execution of green HRM practices. Green HRM encourages organizations to use their current resources consistently by implementing sustainable HRM techniques

[16]. An extensive review of the literature on AI-supported green HRM practices for sustainable and socially responsible growth is what this study aims to do. This research contributes to the development of strategies that support the creation of a greener approach through green HR operations, with a focus on the financial cooperation of forging a robust social responsibility from the perspectives of the community and the company [17]. The study also provides information on how to use technology to change efficient HR into long-term HR. This study helps us to understand:

- The intervention of technologies in the green HRM and their impact on the organization and to build energy-efficient systems.
- Whether it is useful for fast screening and to increase productivity as well as reduce labor cost, without hurting the environment.

This paper is structured as follows: Section 11.2 gives an overview of technology, Section 11.3 discusses the role of AI in green HRM, Section 11.4 describes the challenges, Section 11.5 discusses the suggestions, and Section 11.6 helps us to conclude this study.

11.2 OVERVIEW OF TECHNOLOGY

Now let us understand the technology we are using to get a brief insight of green HRM.

In 2023, women in Africa can expect to live for around 65.0 years after birth, while men can expect to live for about 61.0 years. In nations with high levels of gender parity and women's empowerment, less than 1% of women and girls reside. Over 90% of the world's female population, or 3.1 billion women and girls, reside in nations with significant gender disparities and a lack of women's empowerment, many of which are in Africa.

11.2.1 AI and Women's Empowerment

To advance women's empowerment in Africa, it is imperative that the gender gap and inequality in the AI ecosystem be addressed. The results draw attention to differences in leadership, entrepreneurship, digital infrastructure, education, and economic involvement. Comprehensive policies that place a high priority on inclusive digital infrastructure, skill development, investment for women-led companies, and incentives for women in STEM and tech ecosystems are required to fully realize AI's potential for women's empowerment in Africa. Africa can guarantee a more varied and inclusive ecosystem that benefits society overall by reducing the gender gap and encouraging women to hold leadership positions in AI and technology. For developed or emerging economies, AI is no longer a novel idea. Moreover, intelligent machine creation, especially with regard to computer programs, is the focus of AI science and engineering [18,19,20]. Chaudhary's prediction is that organizations will leverage big data accessibility to enhance their operations through the application of emerging technologies like AI. AI helps many businesses become more productive [21]. Smart machines need to acquire knowledge about human thought and emotion to learn faster. However, with its innovative technological approach, AI has transformed the nature of work and the capacity for decision-making for numerous organizations. Examples of this include genetic algorithms, neural networks, data mining, text mining, sentiment analysis, and interactive voice recognition apps [22]. AI has also been explained in terms of the development of technology. Machine learning and perception technologies are further platforms for technological AI. Machine learning technologies (MLT) concentrate on mimicking human abilities and thought processes to produce more accurate analytical outcomes. It assists in automatically increasing performance and efficiency for a certain task by looking at the pertinent data [23].

AI is defined as "a perfect 'insightful' computer that is an adaptable sane specialist that understands its condition and makes actions that magnify its chance of success at some target." AI can reduce the amount of work that employees have to do by facilitating processes, analyzing data, and other things [24, 25,26]. AI technology links people and machines together as well as between different kinds of machines. The AI framework can be used for candidate screening, employee commitment, re-commitment, career development, and other purposes [27].

Companies are using AI to carry out many tasks more successfully and efficiently. Companies have realized the need of adopting new technology in using their human resource system as a result of the growing pressure to involve HR managers in important decisions [28]. As AI is a new concept in information technology, in the same way the concept of "green human resources" has also recently come to light as a means of conserving natural resources, assuring environmental sustainability, and boosting productivity and corporate growth [29]. Additionally, the study demonstrated that the study demonstrated that, as opportunities to make the industry more sustainable multiply, AI-enabled green HRM provides concrete pathways to reduce environmental impact while enhancing productivity and corporate growth.

Despite being friendly and appealing to women, the data shows that African tech is still a male-dominated field, and significant effort must be made to bring women closer to leadership equality in the industry.

11.3 ARTIFICIAL INTELLIGENCE IN RELATING TO GREEN HRM

Businesses, whether in developing or developed economies, have already embraced a number of contemporary technology to enhance their operations and procedures. Organizations currently frequently use Big Data analytics to examine the trends and patterns in hiring, sales, and production, but applying AI to GHRM practices is still in its infancy [30]. AI has emerged as the key to achieving a lasting competitive edge, but it is still used only by a select few businesses. On the other hand, AI increases productivity and decreases labor costs, which can be better applied to achieving green practices. The idea of AI is novel in the area of internet-based technology, just as the idea of "green human resources" has evolved to protect the environment and ensure its sustainability while also fostering economic growth and productivity [31].

AI has greatly benefited human resource management, or HRM, in every aspect of society. As HRM has advanced in many sectors, it is now also helping to reduce societal pollution. An app called "AskDexter" was recently developed by a Bengaluru-based IT services company. It oversees 22,000 employees by taking care of

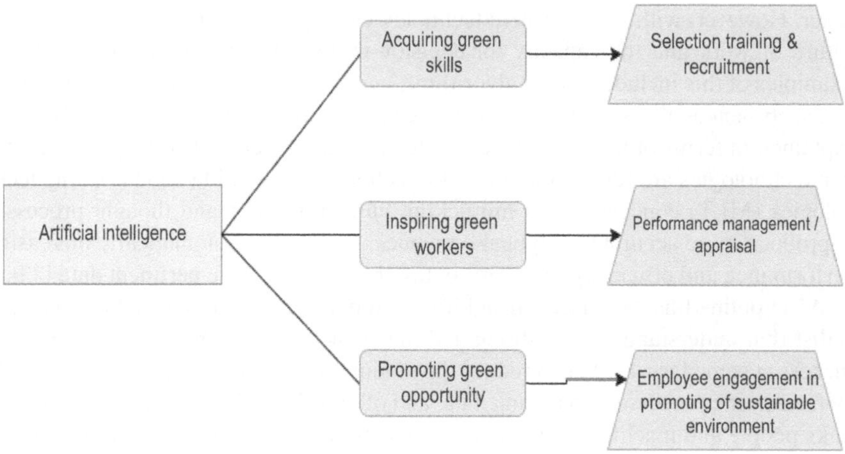

FIGURE 11.2 Artificial intelligence for women empowerment.

HR duties like responding to questions from staff members concerning vacation and corporate policies [32, 33,34]. It responds to requests for technical support and offers straightforward techniques for recognizing employees at all organizational levels. It is capable of providing an option of the desired careers that will be generated in the future, along with the preferred abilities for each [35].

HRM procedures and practices have been dramatically changed by AI and some of fields are shown in Figure 11.2. Green HRM is essential for combining corporate environmental management with human resource management in the modern business world as every organization strives to build innovative, environmentally responsible businesses [36, 37]. The development of AI has resulted in increased business and productivity, environmental sustainability, and protection of natural resources. In this sense, AI is an essential part of HR technology because it can either supplement or improve the efficacy of green HR practice [38]. AI development has previously been hampered by green HRM. The author ends by stating that while speed and agility dominated the first big data phase, the second will concentrate on leveraging machine learning and AI to create significant commercial value [39].

AI helps organizations in developing the green ability of their employees by offering applications that are useful in recruiting and selection, training and development, and identifying candidates as per the job description [40]. AI tools evaluate and choose the best resumes from job applicants that fit the organization's needs. AI apps have reduced the labor of recruiters by up to 60% by leveraging automated and digitalized technology. In the future, 80% of recruiters' work will be eliminated by AI using HR help desk software. Moreover, AI apps assist in lowering the cost of travel for interviewers and recruiters since they are able to conduct interviews using websites like Zoom and Microsoft Teams. Businesses also use AI tools to train their staff, such as Google Classroom and Google Meet platforms for online training [41].

By utilizing performance evaluation tools and data analytics, AI applications assist businesses in inspiring their workforce to adopt green practices. Businesses often evaluate the performance of their employees, giving them feedback for self-evaluation and assistance in addressing their flaws [42]. Employers use AI software to monitor employee behaviour and reward those who adopt environmentally friendly practices [43, 44, 45].

Innovation in technologies embedded with AI is helpful in a variety of green HRM applications and directly supports all of the HR practices, a few of which are highlighted here:

Hiring: AI helps better than humans to analyze and screen CVs based on the profile needed and also saves time.

Performance Evaluation: Modern technologies and data analytics make it possible to evaluate a person's performance on a daily basis as opposed to quarterly or semi-annually. As evaluation is a continual process, it shouldn't be carried out annually [46].

Recruiting and Onboarding: This hiring process works well for the company. AI may help both the employing company and the job seekers during the hiring process. AI has the potential to streamline the application process by producing more intuitive forms that are more likely to be completed by job candidates. This might lead to a significant decrease in the quantity of forms that are returned for review [47].

Operations: A number of chat rooms and message boards are accessible to address a range of staff inquiries. In addition, a number of apps have been created to provide both employees and customers with quick answers. Nowadays, almost all aspects of HR operations contain technology-managed components.

Workforce: A variety of IT organizations are continuously on the lookout for innovation and creativity, but as long as they don't put restrictions on it, businesses will continue to hire creative people, posing a danger to the market leaders.

Green Job Analysis and Design: In this aspect, it is simpler for businesses to formalize green practices while still holding staff members accountable for them. According to AMO theory, a supportive environment is required for employees to perform effectively since they must be given the chance to do so [48].

Green Training and Development: It alludes to teaching employees how to safeguard the environment at work and minimize or prevent environmental pollution. Eco-friendly methods and technologies are frequently employed in training and development initiatives. The AMO theory states that a worker with the requisite skills and knowledge will most likely outperform a worker without them [49].

Green Rewards Management: Employees who exhibit green behavior should be given both monetary and non-monetary awards to effectively meet the organization's green goals. Green rewards management is essential for fostering a green culture among employees [50].

According to the most recent report, green HRM calculates more than 20% of HR compliance, and it has been determined that employing technology reduces the likelihood of error to a minimum while increasing the percentage of human error [51].

To choose and keep green talent, businesses are utilizing automation, such as data mining approaches. Green workers are capable of carrying out duties in accordance with environmental standards; they can be empowered to come up with original solutions for preserving the environment and inspired to protect the environment [52]. Businesses use AI tools like free digital libraries to explore information to improve learning and growth among their employees [53].

11.4 CHALLENGES

While AI brings numerous benefits to HRM, it also poses several challenges that organizations need to address. Here are some key challenges associated with the implementation of AI in HRM:

1. *Bias and Fairness:* AI systems may inherit biases present in historical data, leading to biased decision-making in areas such as recruitment, performance evaluations, and promotions. Ensuring fairness and mitigating biases remains a significant challenge for HR professionals. Women in Africa encounter obstacles to economic participation, with fewer than two-thirds of prime working-age women engaged in the labor force, earning 30–40% less than their male counterparts.
2. *Data Privacy and Security:* In sub-Saharan Africa, there are disparities in digital literacy and educational access, with girls enrolling at lower rates than boys. Female students' exposure to AI education is limited because just 30% of them undertake STEM-related higher education programs. The use of AI involves handling vast amounts of sensitive employee data. Ensuring the privacy and security of this data is critical, and organizations must comply with data protection regulations to prevent unauthorized access or misuse.
3. *Lack of Transparency:* Disparities still exist despite the fact that infrastructure access is essential for women's empowerment in the AI ecosystem. Women in low-income African nations have limited access to AI technologies because just 19% of them have internet access. Many AI algorithms operate as "black boxes," making it challenging for HR professionals and employees to understand the decision-making process. A lack of transparency can lead to distrust and hinder acceptance of AI-based HR practices.
4. *Employee Resistance and Trust Issues:* Only 2% of all venture capital funding goes to women-led firms, making it difficult for women to obtain funding for their businesses. Despite this, firms started by women do better than those founded by men. Despite making up more than half of the population, women in Africa receive only 3% of venture capital funding. Employees may be skeptical or resistant to the use of AI in HRM, fearing job displacement, loss of privacy, or concerns about the fairness of

AI-driven decisions. Building trust and fostering acceptance are ongoing challenges.

5. *Skill Gaps and Training Needs:* Globally, women are underrepresented in AI and scientific R&D roles. As of 2023, women make up only 30% of tech sector professionals in Africa. Women are still underrepresented in leadership positions in the AI and technology industries, despite advancements in encouraging women to hold positions of leadership, especially in politics. The successful integration of AI into HRM requires employees and HR professionals to acquire new skills related to data analytics, AI technology, and ethical considerations. Addressing skill gaps and providing adequate training is crucial.
6. *Legal and Ethical Considerations:* Even though seven African nations have released thorough AI plans, foreign AI technologies frequently control the market, underscoring the necessity of laws that place a higher priority on the development of inclusive digital infrastructure and skills. Legal and ethical issues surrounding the use of AI in HRM, such as compliance with labor laws, anti-discrimination regulations, and maintaining ethical standards, pose challenges for organizations navigating this rapidly evolving field.
7. *Overreliance on Technology:* Relying too heavily on AI without human oversight can lead to suboptimal decision-making. Human judgment and intuition are essential for interpreting context, understanding nuance, and making subjective assessments that AI may struggle with.
8. *Integration with Existing Systems:* Integrating AI systems with existing HRM software and processes can be complex. Ensuring seamless integration, data compatibility, and interoperability with other HR tools is a challenge that organizations may face.
9. *Costs of Implementation:* The initial costs associated with implementing AI in HRM, including acquiring technology, training personnel, and ensuring compliance, can be significant. Organizations need to carefully assess the return on investment and long-term sustainability.
10. *Oversimplification of Complex Issues:* AI algorithms may oversimplify complex human behaviors and interactions. HR processes involve intricate social and emotional dynamics that may not be fully captured by AI, leading to potential misunderstandings or misinterpretations.
11. *Continuous Monitoring and Evaluation:* The need for continuous monitoring and evaluation of AI systems to ensure their ongoing effectiveness and relevance can be resource-intensive. Regular updates and adaptations are necessary to keep pace with evolving organizational needs and changing workforce dynamics.

Addressing these challenges requires a thoughtful and strategic approach, involving collaboration between HR professionals, IT specialists, legal experts, and other stakeholders. Organizations must prioritize ethical considerations, transparency, and continuous learning to maximize the benefits of AI in HRM while minimizing potential risks.

11.5 SUGGESTIONS

Implementing AI in HRM can bring about significant improvements in efficiency, decision-making, and overall workforce management. Here are some suggestions and recommendations for successfully incorporating AI into HRM:

1. *Define Clear Objectives:* Clearly articulate the specific goals and objectives you aim to achieve through the implementation of AI in HRM. Whether it's improving recruitment efficiency, enhancing employee engagement, or optimizing performance evaluations, having well-defined objectives will guide your AI initiatives.
2. *Understand Ethical Implications:* Prioritize ethical considerations and ensure that AI applications align with fairness, transparency, and data privacy standards. Regularly assess and mitigate biases in AI algorithms to avoid unintended consequences.
3. *Employee Involvement and Communication:* Involve employees in the decision-making process and communicate transparently about the introduction of AI in HRM. Address concerns, provide clear explanations of how AI will be used, and emphasize the ways it can benefit employees.
4. *Invest in Employee Training:* Provide training programs for HR professionals and employees to enhance their understanding of AI, its applications, and its impact on HRM. Equip staff with the necessary skills to collaborate effectively with AI technologies.
5. *Start with Pilot Programs:* Begin with small-scale pilot programs to test AI applications in specific HR functions. This allows for the identification of challenges, fine-tuning of algorithms, and the collection of feedback before full-scale implementation.
6. *Ensure Data Accuracy and Quality:* High-quality data is essential for the success of AI applications. Regularly clean and update data to improve accuracy and establish protocols for data governance to maintain the integrity of HR datasets.
7. *Balance Automation with Human Touch:* While automating routine tasks with AI, retain a balance by preserving human involvement in decision-making processes, especially in areas that require emotional intelligence, empathy, and complex judgment.
8. *Integrate AI with Existing Systems:* Ensure seamless integration of AI solutions with existing HRM systems to avoid disruptions and enhance the overall efficiency of HR processes. Compatibility with other tools and technologies is crucial for a cohesive HR ecosystem.
9. *Continuous Monitoring and Evaluation:* Implement regular monitoring and evaluation processes to assess the effectiveness of AI applications in achieving HR goals. Adapt and update AI algorithms based on changing organizational needs and evolving workforce dynamics.
10. *Cybersecurity Measures:* Prioritize cybersecurity to protect sensitive HR data. Implement robust security measures, encryption protocols, and access controls to safeguard AI-driven HR systems from potential threats.

11. *Collaborate with AI Experts:* Collaborate with AI specialists, data scientists, and technology experts to ensure a deep understanding of AI capabilities and limitations. Leverage their expertise to design and implement effective AI solutions.
12. *Stay Informed on Legal Compliance:* Stay updated on legal and regulatory requirements related to AI in HRM. Ensure compliance with labor laws, data protection regulations, and other relevant standards to mitigate legal risks.
13. *User-Friendly Interfaces:* Design user-friendly interfaces for AI applications in HRM to encourage adoption and acceptance. Ease of use and accessibility will contribute to positive user experiences among HR professionals and employees.

By following these recommendations, organizations can navigate the integration of AI into HRM more effectively, promoting positive outcomes for both the workforce and the organization as a whole.

There are some suggestions for the business that can use its advantages and turn its weaknesses into opportunities for organizations.

- First and foremost, the organizations need to strengthen its human resource management, particularly in terms of hiring and performance monitoring using AI. Developing fundamental and advanced digital skills that are suited to women's interests and demands is crucial, as is providing capital, networks, and market access to digitally enabled businesses.
- Reducing the digital divide can also be achieved by enhancing the accessibility of goods and services, particularly government services, through digital identity (ID).
- The organization should increase staffing in the green human resources division and give them professional development opportunities. Organization should found ways with customers and formulate ways on how to enhance their sales from green practices.
- Promoting greater female leadership in policymaking is essential to ensuring that African parliaments produce better laws and regulations. Establishing gender-sensitive laws that support equal opportunities, access, and compensation for all can contribute to the development of a more welcoming workplace. More women can enter and succeed in the computer industry if laws and regulations are changed to offer advantages to women in the sector, such as tax breaks and reduced regulatory rates.
- Organizations need to investigate cost-effective ways to boost their bottom line while employing AI to fund their green initiatives. In essence, by using AI, GHRM would guide the business to better success in the future.

11.6 CONCLUSION

The Maputo Protocol on Women's Rights has been ratified by 44 of the 55 African Union Member States as of June 2023, which is encouraging. To promote gender equality and guarantee that women's rights are upheld in the face of technical breakthroughs like AI, this protocol is essential. African nations can strive toward a more inclusive and equitable future by coordinating AI policy with current frameworks for gender equality. The information regarding AI-supported green HRM practices for social responsibility and environmental sustainability is summarized in this study. To maintain a competitive edge in the business sector, organizations from all over the world are attempting to adopt GHRM concepts. Full acceptance and adoption of GHRM in the firm are not difficult to achieve, but it does require a change in management and employee attitudes about current HR practices. The idea of GHRM has been developed to safeguard environmental resources. This might mean that environmental sustainability is ensured while simultaneously growing the business and boosting productivity. AI supports the adoption of GHRM services by providing a range of applications and intelligent technologies that streamline tasks and boost worker productivity. According to earlier research, the key elements for implementing GHRM practices are training, recruitment, and rewards. So these procedures can make use of AI tools and approaches to engage staff and encourage voluntary participation in environmentally friendly activities. Using GHRM practices improves environmental performance and organizational sustainability. In-depth explanations of the factors that influence GHRM practices make it easier to pinpoint the appropriate AI applications for them. Earlier studies also recommended that the utilization of AI, Big Data analytics, and people/HR analytics should be taken into consideration for effective adoption of e-HRM practices. Further studies can be done to examine the implications of how AI affects the applications of GHRM practices and to investigate the specialty areas of GHRM research.

In conclusion, to advance women's empowerment in Africa, it is imperative that the gender gap and inequality in the AI ecosystem be addressed. The results draw attention to differences in leadership, entrepreneurship, digital infrastructure, education, and economic involvement. Comprehensive policies that place a high priority on inclusive digital infrastructure, skill development, investment for women-led companies, and incentives for women in STEM and tech ecosystems are required to fully realize AI's potential for women's empowerment in Africa. Africa can guarantee a more varied and inclusive ecosystem that benefits society overall by reducing the gender gap and encouraging women to hold leadership positions in AI and technology.

The integration of AI into HRM marks a paradigm shift in how organizations approach workforce management. The advent of AI brings forth the promise of increased efficiency, data-driven decision-making, and enhanced employee experiences. As organizations leverage AI tools for recruitment, talent management, and performance analysis, they stand to gain a competitive edge in the rapidly evolving business landscape. However, this transformative journey is not without challenges, with ethical considerations, data privacy concerns, and the need for transparent communication posing critical hurdles. Striking a delicate balance between technological innovations and preserving the human touch in HR processes is key to success. As AI continues to evolve, organizations that navigate these challenges with strategic foresight, ethical integrity, and a commitment to employee well-being are likely to unlock the full potential of AI in reshaping the future of HRM. The journey ahead involves not only embracing the capabilities of AI but also fostering a culture that values continuous learning, collaboration, and a harmonious blend of human and AI for organizational prosperity.

REFERENCES

[1] Zubair, D. S. S., & Khan, M. (2019). Sustainable development: The role of green HRM. *International Journal of Research in Human Resource Management, 1*(2), 1–6.
[2] Garg, V., Srivastav, S., & Gupta, A. (2018, October). Application of artificial intelligence for sustaining green Human Resource Management. In *2018 International Conference on Automation and Computational Engineering (ICACE)* (pp. 113–116). IEEE.
[3] Bijlwan, S., Kaur, M., Sharma, A., Singh, T., & Saini, N. (2024). Impact of technology on healthcare along with social security to achieve sustainability. In *Sustainable Technology for Society 5.0* (pp. 12–18). CRC Press.
[4] Sekhri, A., & Cheema, D. J. (2019). The new era of HRM: AI reinventing HRM functions. *International Journal of Scientific Research and Review, 7*(3), 3073–3077.
[5] Ahmad, S. (2015). Green Human Resource Management: Policies and practices. *Cogent Business & Management, 2*(1), 1030817.
[6] Garg, V., Srivastav, S., & Gupta, A. (2018, October). Application of artificial intelligence for sustaining green Human Resource Management. In *2018 International Conference on Automation and Computational Engineering (ICACE)* (pp. 113–116). IEEE.
[7] Fachada, J., Rebelo, T., Lourenço, P., Dimas, I., & Martins, H. (2022). Green Human Resource Management: A bibliometric analysis. *Administrative Sciences, 12*(3), 95.
[8] Shukla, K., & Bankar, S. (2022). Adoption of green HRM practices for building sustainable models in the Indian markets. *SDMIMD Journal of Management, 13*(1).
[9] Bekhit, K., El Leithy, W., & Mahmoud, A. (2023). The Impact of Green Human Resource Management Practices on Employee Retention and Environmental Sustainability: A conceptual model. *22*, 1005–1034. https://doi.org/10.37896/YMER22.01/78
[10] Pham, N. T., Hoang, H. T., & Phan, Q. P. T. (2020). Green Human Resource Management: A comprehensive review and future research agenda. *International Journal of Manpower, 41*(7), 845–878.
[11] Iddagoda, A., Dissanyake, H., & Abeysinghe, R. (2022). Green Human Resource Management: A bibliometric analysis. *Romanian Journal of Economics, 55*(2 (64)), 147–159.
[12] Hossin, M. S., Ulfy, M. A., & Karim, M. W. (2021). Challenges in adopting artificial intelligence (AI) in HRM practices: A study on Bangladesh perspective. *International Fellowship Journal of Interdisciplinary ResearchVolume, 1.*
[13] Gupta, M. (2021). The role of artificial intelligence in adopting green HRM practices. *Reinventing Manufacturing and Business Processes Through Artificial Intelligence*, 1–18.
[14] Bijlwan, S., & Srivastava, V. P. (2022). A review of E-training and development practices implemented by government for healthcare sector and their impact during COVID 19 (with special reference to Uttarakhand). *Specialusis Ugdymas, 1*(43), 3976–3988.
[15] Renwick, D. W., Redman, T., & Maguire, S. (2013). Green Human Resource Management: A review and research agenda. *International Journal of Management Reviews, 15*(1), 1–14.
[16] Garg, V., Srivastav, S., & Gupta, A. (2018, October). Application of artificial intelligence for sustaining green Human Resource Management. In *2018 International Conference on Automation and Computational Engineering (ICACE)* (pp. 113–116). IEEE.
[17] Das, S. C., & Singh, R. K. (2016). Green HRM and organizational sustainability: An empirical review. *Kegees Journal of Social Science, 8*(1), 227–236.
[18] Amrutha, V. N., & Geetha, S. N. (2020). A systematic review on green Human Resource Management: Implications for social sustainability. *Journal of Cleaner Production, 247*, 119131.

[19] Sharma, R., & Gupta, N. (2015, January). Green HRM: An innovative approach to environmental sustainability. In *Proceeding of the Twelfth AIMS International Conference on Management* (pp. 2–5).

[20] Ullah, M. M. (2017). Integrating environmental sustainability into human resources management: A comprehensive review on green human resources management (Green HRM) practices. *Economics and Management*, *6*(1), 14–19.

[21] Yong, J. Y., Yusliza, M. Y., Ramayah, T., Chiappetta Jabbour, C. J., Sehnem, S., & Mani, V. (2020). Pathways towards sustainability in manufacturing organizations: Empirical evidence on the role of green Human Resource Management. *Business Strategy and the Environment*, *29*(1), 212–228.

[22] Abbas, Z., Sarwar, S., Rehman, M. A., Zámečník, R., & Shoaib, M. (2022). Green HRM promotes higher education sustainability: A mediated-moderated analysis. *International Journal of Manpower*, *43*(3), 827–843.

[23] Rani, S., & Mishra, K. (2014). Green HRM: Practices and strategic implementation in the organizations. *International Journal on Recent and Innovation Trends in Computing and Communication*, *2*(11), 3633–3639.

[24] Gholami, H., Rezaei, G., Saman, M. Z. M., Sharif, S., & Zakuan, N. (2016). State-of-the-art green HRM system: Sustainability in the sports center in Malaysia using a multi-methods approach and opportunities for future research. *Journal of Cleaner Production*, *124*, 142–163.

[25] Emel, E. S. E. N., & Caliskan, A. O. (2019). Green Human Resource Management (GHRM) and environmental sustainability. *PressAcademia Procedia*, *9*(1), 58–60.

[26] Amjad, F., Abbas, W., Zia-Ur-Rehman, M., Baig, S. A., Hashim, M., Khan, A., & Rehman, H. U. (2021). Effect of green Human Resource Management practices on organizational sustainability: The mediating role of environmental and employee performance. *Environmental Science and Pollution Research*, *28*, 28191–28206.

[27] Yong, J. Y., Yusliza, M. Y., & Ramayah, T., & Khalid, F., & Opel, M. I. (2022). Accentuating the interconnection between Green Intellectual Capital, Green Human Resource Management and Sustainability. *Benchmarking: An International Journal*. *30*. https://doi.org/10.1108/BIJ-11-2021-0641.

[28] Deshwal, P. (2015). Green HRM: An organizational strategy of greening people. *International Journal of Applied Research*, *1*(13), 176–181.

[29] Sathasivam, K., Abu Bakar, R., & Che Hashim, R. (2021). Embracing organisational environmental sustainability: Experiences in green Human Resource Management. *Business Strategy & Development*, *4*(2), 123–135.

[30] Chaudhary, R. (2020). Green Human Resource Management and employee green behavior: An empirical analysis. *Corporate Social Responsibility and Environmental Management*, *27*(2), 630–641.

[31] Juyal, A., & Baijwan, S. (2023, February). An improved analysis of tourism management in enhanced forest environment using block chain based internet of things model. In *2023 IEEE International Conference on Integrated Circuits and Communication Systems (ICICACS)* (pp. 1–6). IEEE.

[32] Juyal, A., & Baijwan, S. (2023, February). An improved analysis of tourism management in enhanced forest environment using block chain based internet of things model. In *2023 IEEE International Conference on Integrated Circuits and Communication Systems (ICICACS)* (pp. 1–6). IEEE.

[33] Amjad, F., Abbas, W., Zia-Ur-Rehman, M., Baig, S. A., Hashim, M., Khan, A., & Rehman, H. U. (2021). Effect of green Human Resource Management practices on organizational sustainability: The mediating role of environmental and employee performance. *Environmental Science and Pollution Research*, *28*, 28191–28206.

[34] Alghamdi, S. (2021). Effect of green human resource practices on the employee performance and behavior: A Systematic. *International Journal of Economics, Business and Management Research*, *5*(3), 450–464.

[35] Shahzad, M. A., Jianguo, D., & Junaid, M. (2023). Impact of green HRM practices on sustainable performance: Mediating role of green innovation, green culture, and green employees' behavior. *Environmental Science and Pollution Research*, *30*(38), 88524–88547.

[36] Khan, M. H., & Muktar, S. N. (2024). Green employee empowerment: The missing linchpin between green HRM and sustainable organizational performance. *Journal of Cleaner Production*, *434*, 139812.

[37] Gill, A., Ahmad, B., & Kazmi, S. (2021). The effect of green Human Resource Management on environmental performance: The mediating role of employee eco-friendly behavior. *Management Science Letters*, *11*(6), 1725–1736.

[38] Rubel, M. R. B., Kee, D. M. H., & Rimi, N. N. (2021). The influence of green HRM practices on green service behaviors: The mediating effect of green knowledge sharing. *Employee Relations: The International Journal*, *43*(5), 996–1015.

[39] Ojo, A. O., Tan, C. N. L., & Alias, M. (2022). Linking green HRM practices to environmental performance through pro-environment behaviour in the information technology sector. *Social Responsibility Journal*, *18*(1), 1–18.

[40] Muisyo, P. K., & Qin, S. (2021). Enhancing the FIRM'S green performance through green HRM: The moderating role of green innovation culture. *Journal of cleaner production*, *289*, 125720.

[41] Wulansari, N. A., Witiastuti, R. S., & Ridloah, S. (2018). Employee performance measurement development based on Green HRM indicators. *KnE Social Sciences*, 1179–1194.

[42] Rani, S., & Mishra, K. (2014). Green HRM: Practices and strategic implementation in the organizations. *International Journal on Recent and Innovation Trends in Computing and Communication*, *2*(11), 3633–3639.

[43] Jeronimo, H. M., de Lacerda, T. C., & Henriques, P. L. (2020). From sustainable HRM to employee performance: A complex and intertwined road. *European Management Review*, *17*(4), 871–884.

[44] Nisar, Q. A., Haider, S., Ali, F., Gill, S. S., & Waqas, A. (2022). The role of green HRM on environmental performance of hotels: Mediating effect of green self-efficacy & employee green behaviors. *Journal of Quality Assurance in Hospitality & Tourism*, 1–34.

[45] Suharti, L., & Sugiarto, A. (2020). A qualitative study of green HRM practices and their benefits in the organization: An Indonesian company experience. *Business: Theory and Practice*, *21*(1), 200–211.

[46] He, J., Morrison, A. M., & Zhang, H. (2021). Being sustainable: The three way interactive effects of CSR, green Human Resource Management, and responsible leadership on employee green behavior and task performance. *Corporate Social Responsibility and Environmental Management*, *28*(3), 1043–1054.

[47] Tahir, M., Safwan, N., Usman, A., & Adnan, A. (2020). Green HRM as predictor of firms' environmental performance and role of employees' environmental organizational citizenship behavior as a mediator. *Journal of Research and Reviews in Social Sciences Pakistan*, *3*(1), 699–715.

[48] Jayabalan, N., Zafir, M. M., Kumar, R. M. H. M., Hayati, Y., & Mai, F. M. B. (2020). The role of OCBE on green HRM towards performance sustainability. *International Journal of Innovation, Creativity and Change*, *13*(5), 388-399.

[49] Suba, M., Inayath Ahamed, S. B., & Priya, B. S. (2021). Green-HRM a strategic move towards the CSR through green employee performance. *M. Suba, SB Inayath*

Ahamed and B. Shanmuga Priya, Green-HRM A Strategic Move Towards the CSR Through Green Employee Performance, International Journal of Management, 11(11), 2020.

[50] Chuah, S. C., Mohd, I. H., Kamaruddin, J. N. B., & Md Noh, N. (2021). Impact of Green Human Resource Management practices towards green lifestyle and job performance. *Global Business & Management Research, 13*(4).

[51] Muisyo, P., Su, Q., Ho, T. H., Julius, M. M., & Usmani, M. S. (2022). Implications of green HRM on the firm's green competitive advantage: The mediating role of enablers of green culture. *Journal of Manufacturing Technology Management, 33*(2), 308–333.

[52] Uddin, M. M., & Islam, R. (2015). Green HRM: Goal attainment through environmental sustainability. *Journal of Nepalese Business Studies, 9*(1), 14–19.

12 Analyzing the Marketing Challenges of Sustainable Products among Women Entrepreneurs in Nigeria

Anuja Shukla, Rotimi Adeniran Owoade, and Aissam Bouadjel

12.1 INTRODUCTION

In the recent past, the global quest for sustainability has gained momentum, leading to increased production and sale of sustainable products. This is very vital in the emerging economies such as Nigeria, where sustainable practices are being viewed as opportunities for environmental conservation and economic growth. Women entrepreneurs are part of the leading drivers of the movement, and they are noted to be leading the way in driving sustainable development. Even though they are endowed with potential, such entrepreneurs are faced with multidimensional marketing challenges that limit their capacity to reach the market beneficially with sustainable products (Walden University, 2023; ResearchGate, 2023).

Nigerian female entrepreneurs are facing marketing challenges squarely based on socioeconomic and cultural factors. Insufficient funds and restricted access to funds are prevalent, discouraging them from investing in marketing strategies that will keep their products in the spotlight in the competitive market (Tandfonline, 2022). Moreover, the lack of infrastructure and market data is the primary challenge, given that most women entrepreneurs are not revealed and available to a vast population of customers (Munich Personal RePEc Archive, 2019).

Additionally, gender discrimination continues to exist, denying women entrepreneurs access to networks and resources that are generally more available to their male counterparts (EUDL, 2021). This gender inequality is also further aggravated by work–family conflicts and societal expectations, which are more prone to place further burdens on women entrepreneurs, preventing them from giving their best to business undertakings (Women's World Banking, 2024).

Even with such challenges, the growth potential for the Nigerian women entrepreneurs' sustainable products market is high. By overcoming such challenges, the

DOI: 10.1201/9781003508953-12

entrepreneurial environment can be enhanced, with sustainable economic growth of the region being guaranteed (ResearchGate, 2023).

12.2 LITERATURE REVIEW

Many studies project financial limitations as a core obstacle for women entrepreneurs selling sustainable goods. Based on a study undertaken by Walden University (2023), lack of access to capital seriously stifles the efforts of women entrepreneurs to invest in marketing tactics important for effective competitive market positioning. This capital-related obstacle is augmented by the dearth of collateral security, preventing access to financial resources from official financial institutions (Tandfonline, 2022).

Contrasting literature, nonetheless, identifies the functionality of microfinance as an asymptomatic enabler for women entrepreneurs. Academia.edu published a report highlighting the significant influence of microfinance in empowering women entrepreneurs through provision of the much-needed capital to expand business and enhance market penetration (Academia.edu, 2023).

12.3 CULTURAL AND GENDER-BASED CHALLENGES

Cultural demands and gender discrimination also critically contribute to designing the entrepreneurial process for women in Nigeria. Female entrepreneurs tend to be subject to social pressures exerted by additional family care requirements, which would distract from the focus of the business (Women's World Banking, 2024). ResearchGate (2023) speaks to how the social norms of the environment bar women from obtaining access to business networks and funds traditionally controlled by men.

In contrast, some studies highlight adaptive strategies employed by women entrepreneurs to navigate these cultural barriers. According to a study on ResearchGate (2023), women entrepreneurs are increasingly leveraging digital platforms to circumvent traditional gender roles, thereby expanding their market access and visibility.

12.4 INFRASTRUCTURAL AND MARKET INFORMATION BARRIERS

The absence of infrastructure and market information is still a significant setback. The majority of women entrepreneurs work in low-infrastructure areas, thereby restricting their capacity to enable efficient product distribution (Munich Personal RePEc Archive, 2019). Additionally, a shortage of market information restricts their capacity to comprehend and access potential customers (EUDL, 2021).

Conversely, other perspectives believe that digitalization can actually mitigate these challenges. As quoted in Covenant University (2023), adopting digital platforms and tools provides an opportunity to bridge information gaps and enhance market access for women entrepreneurs.

The background for women's entrepreneurship in Nigeria is one of superior opportunities and superior challenges, most especially in the marketing of sustainable products. There is a consensus in existing literature that there is a complex interaction of socioeconomic, cultural, and infrastructural factors that define the experiences of women entrepreneurs in the industry.

12.5 CHALLENGES IN MARKETING ECOLOGICALLY FRIENDLY PRODUCTS

Nigerian women entrepreneurs have many challenges in marketing sustainable products. One of the biggest challenges is the limitation of financial resources, which restricts them from engaging in good marketing practices. A survey by Walden University (2023) shows that the lack of adequate access to capital is a major challenge that keeps women entrepreneurs from accessing marketing platforms that increase their visibility. In addition, the unavailability of market information and infrastructure is a major challenge that keeps them from marketing. A survey by the Munich Personal RePEc Archive (2019) shows that these entrepreneurs prefer to work in environments where access to market information is significantly restricted, which negatively impacts their ability to make informed decisions.

Gender imbalance is a prevailing trend. Despite their economic input, women business owners are faced with gender inequality that hinders their access to networks and resources (EUDL, 2021). Furthermore, work–family conflicts and societal expectations add to these issues, as observed in a Women's World Banking (2024) study that identifies the added pressures of women in juggling entrepreneurial responsibilities and family obligations.

Contrasting views presented in the literature are that despite such challenges, they are not impossible to overcome. For example, a ResearchGate (2023) study presents the resilience of Nigerian women entrepreneurs and their ability to innovate in limited situations. The study provides examples where women have been able to overcome marketing challenges using social networks and taking advantage of digital platforms to access wider networks. Additionally, others contend that the emphasis on challenges should not be at the expense of the opportunities that exist. The Women Entrepreneurship and Sustainability article (Covenant University, 2023) indicates that the transition toward sustainable products is a novel market opportunity. Through aligning their products with global sustainability objectives, women entrepreneurs can access niche markets that are increasingly attractive to socially responsible consumers.

12.6 POTENTIAL FOR GROWTH

The growth of this industry has a bright prospect, as it has been seen through the number of schemes focused on empowering women business owners in Nigeria. Furthermore, the literature shows that strategic interventions, including greater access to financial services and better market infrastructure, have the ability to substantially reduce the marketing difficulties of these business owners (Tandfonline, 2022). When these constraints are improved upon, there is a possibility of women business owners contributing significantly to sustainable economic growth in Nigeria.

12.7 LIMITED MARKET ACCESS AND FINANCING

Market and finance access presents a significant constraint to women entrepreneurs with sustainable products. Research indicates that women tend to use informal financial networks like *esusu* due to the unavailability of formal financing (Peter & Orser, 2024). Lacking sufficient capital, women entrepreneurs are unable to invest in marketing outlets, branding initiatives, and customer outreach activities essential for sustainable product promotion.

Additionally, Odunaike and Ajiboye (2020) observe that poor women entrepreneurs, especially in border communities, experience further challenges in accessing broader markets, which hinders their capacity to scale sustainable businesses. Their scarce financial means hinder their participation in trade fairs and marketing events where sustainable products may achieve exposure.

12.8 DIGITAL TECHNOLOGY AND MARKETING CHALLENGES

The use of digital technologies is crucial for modern marketing, but women entrepreneurs in Nigeria encounter considerable challenges in this regard. Etim (2021) stresses that although digital tools have the potential to break geographical barriers, limitations like digital illiteracy, expense, and inadequate infrastructure limit the application of e-commerce, social media marketing, and digital branding among women.

Lacking an effective digital presence, women-owned sustainable products are localized and not seen by wider consumer markets that increasingly seek out sustainability.

12.9 CULTURAL AND STRUCTURAL BARRIERS

Cultural expectations weigh heavily on marketing practices. Female gender roles in Nigeria tend to exclude women entrepreneurs from male-dominated markets (Peter & Orser, 2024). The negotiating power of women is curtailed, and it has implications for how they position sustainable products and are able to get competitive prices. The patriarchal institutions and the gender biases weaken the credibility and legitimacy of female-led brands in the market.

This is a problem that aggravates the challenge of marketing products that are already in need of consumer education because of their sustainability.

12.10 ENTREPRENEURIAL EDUCATION AND NETWORKS

Entrepreneurial education and support networks are necessary to address marketing challenges. Etim (2021) points out the importance of education in developing women's digital and marketing skills. Training in sustainable branding, consumer engagement, and market research can empower women to market their products more effectively. Yet the majority of current programs are urban-oriented, with rural women entrepreneurs lacking access to required resources (Sanni, 2024).

12.11 RESEARCH GAP

The literature identifies that Nigerian women entrepreneurs encounter multidimensional marketing challenges in marketing sustainable products. Central hindrances encompass limited market access, limited funds, technological barriers, and cultural prejudices. These challenges require a multifaceted response comprising policy intervention, capacity development, and systemic cultural change. Wider entrepreneurial education and the establishment of inclusive digital infrastructures would have a significant impact on increasing the visibility and success of women-owned sustainable businesses (Table 12.1).

12.12 RESEARCH METHODOLOGY

To solve the aforementioned issues, a qualitative study was conducted. Primary data were collected from 15 Nigerian female entrepreneurs through semi-structured interviews. This approach allowed for the collection of descriptive and intimate histories,

TABLE 12.1
Literature Review Matrix

Concept	Key Authors	Key Findings
Limited market access and financing	Peter and Orser (2024); Odunaike and Ajiboye (2020)	Women entrepreneurs rely on informal savings (*esusu*) due to a lack of formal financing; limited market reach hinders sustainable product visibility.
Digital technology and marketing challenges	Etim (2021)	Digital tools could bridge geographic gaps, but barriers like digital illiteracy, infrastructure issues, and cost limit adoption for marketing.
Cultural and structural constraints	Peter and Orser (2024)	Patriarchal norms and male-dominated markets limit women's negotiation power and reduce credibility of women-led sustainable brands.
Entrepreneurial education and networks	Etim (2021); Sanni (2024)	Entrepreneurial education and network participation improve marketing skills, but rural women entrepreneurs remain underserved by current programs.
Research gaps	Synthesized from all sources	Lack of focus on consumer perceptions, comparative effectiveness of marketing methods (digital vs. traditional), and supportive policy frameworks.

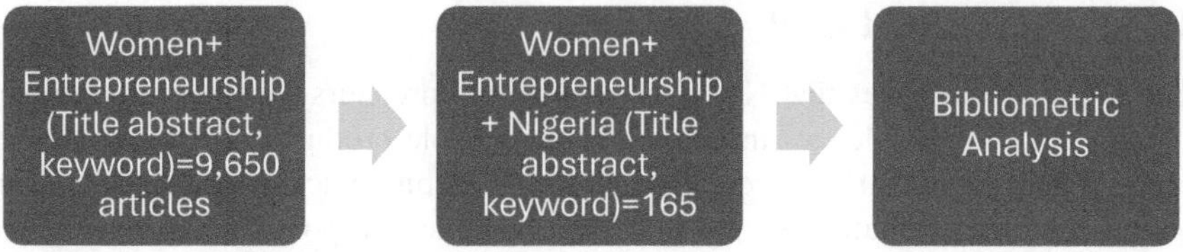

FIGURE 12.1 Data collection process.

allowing detailed investigation of the most critical issues influencing their entrepreneurial ventures. The findings obtained provide valuable insights that quantitative methods only may fail to provide. Figure 12.1 represents the processes followed in data collection for bibliometric analysis. A five-step process was used: Identification of keywords, Search in databases, Screening of documents, Extraction of data, and Analysis. We employed a systematic method: choosing keywords for "Women Entrepreneurship" from search database Scopus leading to extraction of 9,650 articles. The filter was applied with keyword "Nigeria" and language in English leading to extraction of 165 articles.

12.13 RESULTS

The results of the bibliometric analysis are discussed as follows. Figure 12.2 represents the growth of number of articles. Significant growth has been noticed from 2007 onwards in the field of women's entrepreneurship studies in Nigeria. Figure 12.3 explains which authors have contributed most to the academic field of study. The highest number of papers are contributed by Akinbami, followed by Halkias, and Nwajuba. Figure 12.4 shows the number of documents contributed by various institutions (affiliations). The top three contributing institutions are the University of Nigeria, Covenant University, and Obafemi Awolowo University. This chart indicates the universities or institutions that dominate the research in women's entrepreneurship and sustainability in Nigeria. Figure 12.5 indicates the global spread of research contribution. The top contributors are Nigeria, South Africa, the United States, and the United Kingdom. Figure 12.6 represents the composition of articles as per type, while Figure 12.7 represents subject area-wise contribution to the growing field of studies in women's entrepreneurship in Nigeria.

The results of interview are discussed in Table 12.2 and Figure 12.8.

12.14 DISCUSSION

The businesses studied focus heavily on eco-friendly, organic, and sustainable alternatives across various industries. They cover a wide range of sectors, including skincare, fashion, home goods, food packaging, furniture, and even technology. A clear trend emerges where sustainability is not limited to one area but is being integrated into many aspects of lifestyle and daily living.

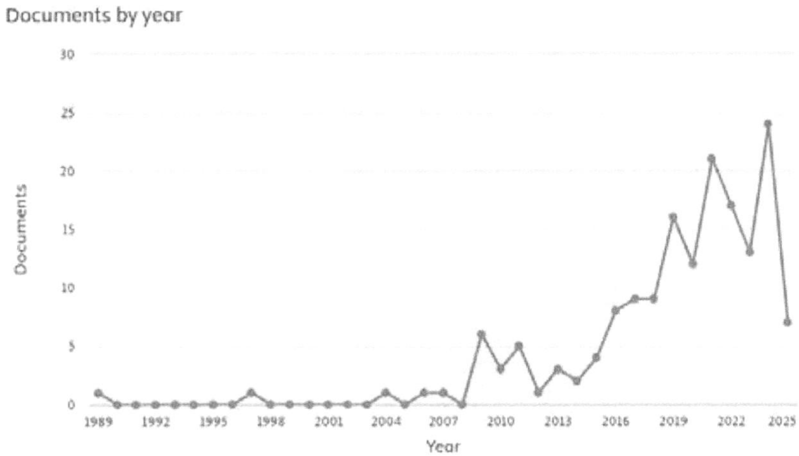

FIGURE 12.2 Publication trend.

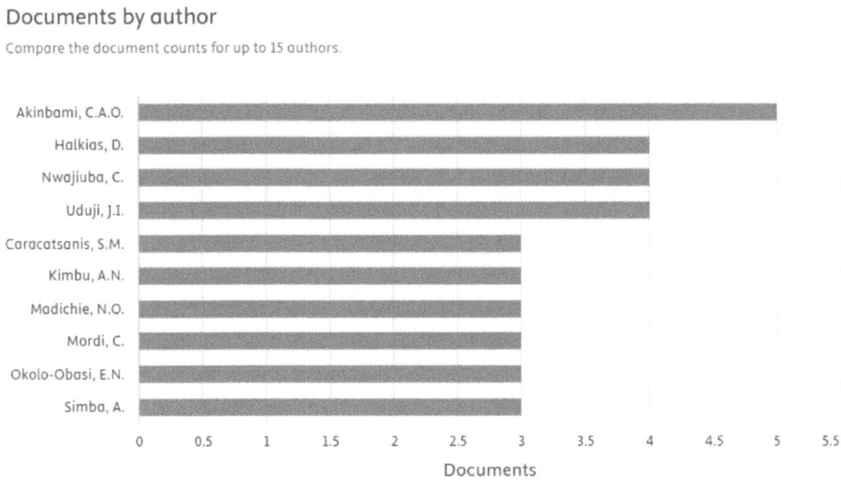

FIGURE 12.3 Documents per author.

The motivations for beginning these businesses are largely personal and extremely values-led. The majority of the entrepreneurs were inspired by their own lives, for instance, overcoming allergies, by a passion for natural hair care, or by a passion for green living. Others spotted glaring market gaps, for instance, the lack of green fashion or green furniture, and others were inspired by bigger causes such as the removal of plastic pollution or the promotion of renewable energy. Overall, it can

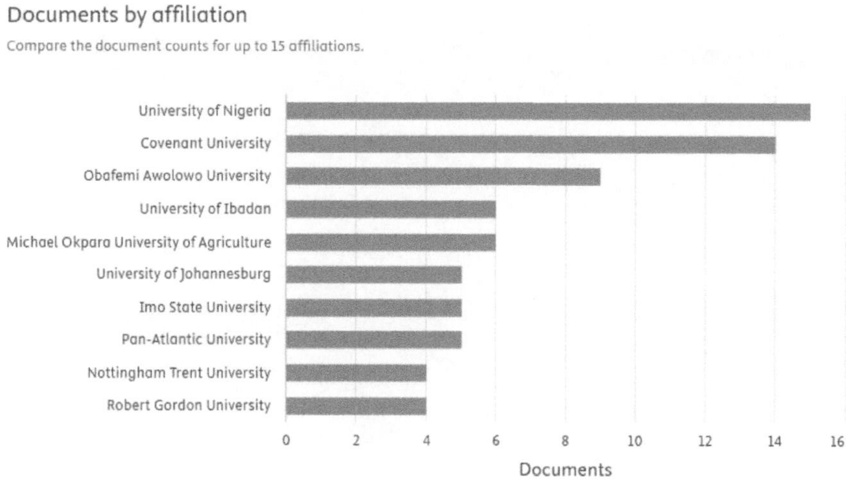

FIGURE 12.4 Documents by affiliation.

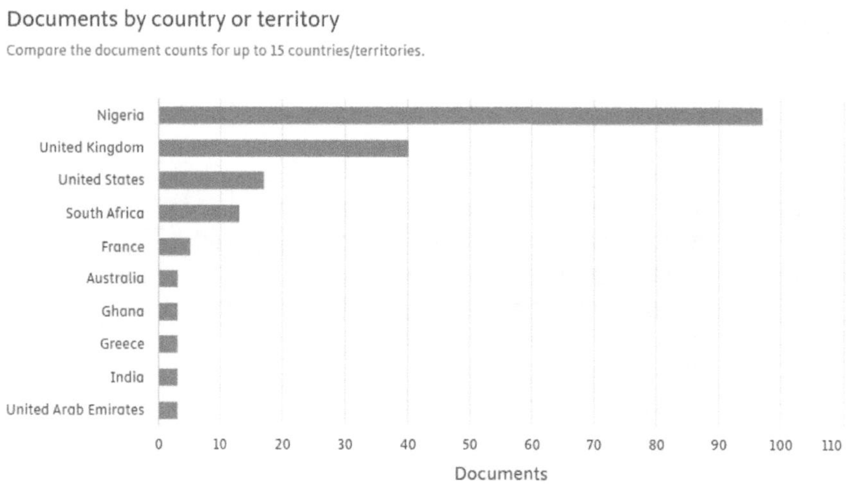

FIGURE 12.5 Documents per country.

be observed that these entrepreneurs are cause-driven, more inspired by making a difference rather than by profit per se.

But it is a challenging task to sell these green products. There are steep prices and perceptions for business owners, and green products are unaffordable to consumers. There is always also the issue of building awareness and trust since most consumers remain skeptical or unaware of the real benefits of green products. Added to this is also rising competition and market saturation, making it

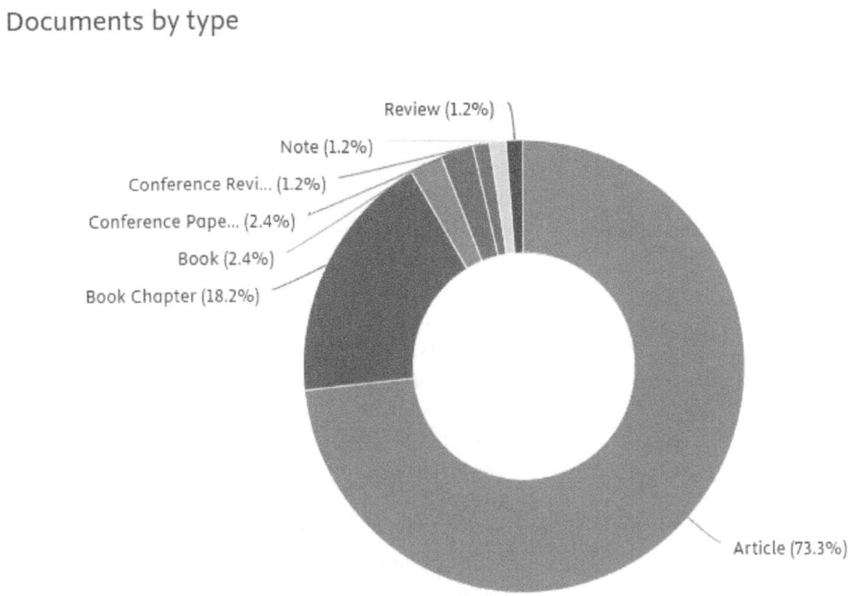

FIGURE 12.6 Composition of articles as per type.

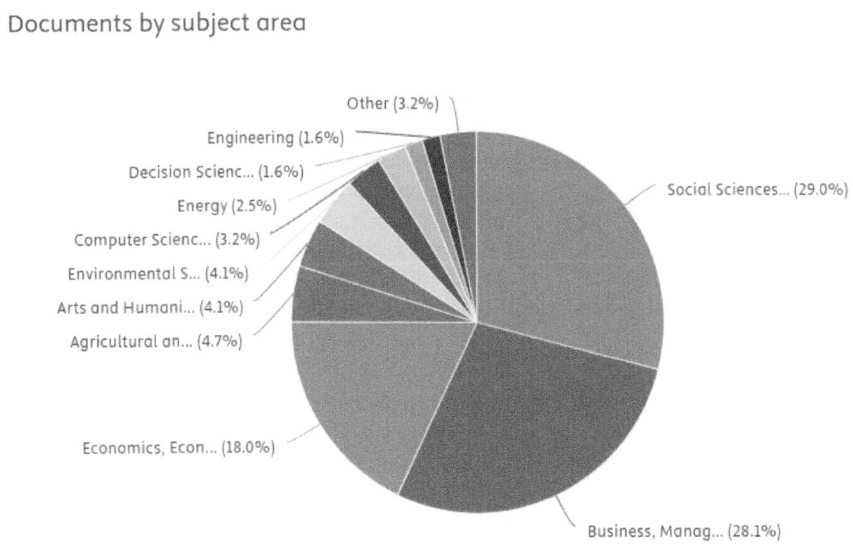

FIGURE 12.7 Documents by subject area.

TABLE 12.2
Qualitative Analysis

How long have you been in business?	**5.6 average (max =10, min 2)**
No. of employees	Average = 8 (max = 15, min = 3)
Nature of business and sustainable products	organic skincare, eco-fashion, organic farm, eco-home goods, recycled handbags, vegan cosmetics, eco-food packaging, handmade soaps, eco-furniture, reusable kitchenware, natural haircare, plant-based foods, biodegradable bags, solar-powered gadgets, eco-cleaning products.
Motivation to start business	Promote healthy skin, fill eco-fashion gap, organic food needs, love for eco-living, reduce plastic waste, personal allergy issues, fight food waste, promote handmade culture, sustainable furniture gap, reduce kitchen waste, natural haircare passion, veganism, fight plastic pollution, renewable energy passion, green cleaning awareness.
Marketing challenges	Price perception, low awareness, funding issues, consumer education, competition, market saturation, lack of trust, awareness building, high costs, pricing struggles, small audience, trust issues, expensive raw materials, customer skepticism, price comparison.
Consumers perceive sustainable products	Expensive, luxury, confusing, gimmicky, necessary but pricey, luxury, trendy, new concept, eco but costly, doubtful, healthy, important, not urgent, too fancy, growing interest
Strategies used for overcoming challenges	Blogging, community fairs, customer education, customer reviews, discounts, eco-ambassador program, eco-campaigns, eco-certifications, influencer marketing, loyalty programs partner collabs, product demos, social media ads, storytelling, workshops
Source of motivation/success stories	Community leader award, Facebook ad success, fair exhibition, influencer boost, Instagram growth, local awards, national feature new export deal, NGO recognition, partnership, podcast guest, radio interview, school workshops, tv feature, viral post
Support needed for market growth	Awareness drives, digital training, eco-festivals, education campaigns, export support, farmer training, local brand support, national campaigns, online presence, partnership initiatives, price subsidy, public events, rural education, youth education, youth programs
Recommendations to policymakers	Digital infrastructure, eco-awareness, eco-policies, free trainings green awards, marketing funding, marketing support, more grants partnership programs, rural focus, SME loans, subsidies, tax reduction, trade shows, women entrepreneur hubs

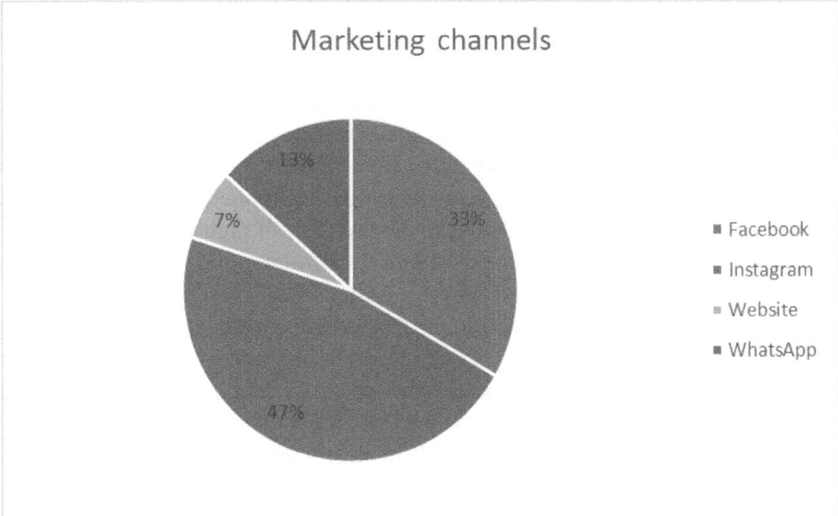

FIGURE 12.8 Marketing channels.

challenging for small brands to break through. Even with the expansion of the green market, there still needs to be a lot of work done in educating consumers and tackling the cost-related issues.

On the consumer attitudes front, the results are mixed. Some consider sustainable products to be fashionable and healthy, while others consider them luxury products that are not necessarily a priority. Most consider the concepts confusing or even gimmicky, and pricing is a frequent problem. While there is increasing interest, the urgency to use eco-products is low. Consumers enjoy the concept of sustainability but are reluctant to make a buying decision.

To overcome such marketing obstacles, companies have turned to creative and grassroots methods. Most of them place great emphasis on education by blogging, workshops, storytelling, and word of mouth. Community outreach is another popularity strategy, employing fairs, green efforts, and ambassador programs to build trust and visibility. Collaborations and partnerships with influencers and targeted social media advertising are also part of their strategies. In general, the top-performing brands are eager to not only sell a product but also to share a story and mission that speaks to consumers.

Success stories demonstrate that external validation is used to encourage entrepreneurs and compel businesses to scale. Validation through awards, media attention, viral social media stories, and collaborations has granted legitimacy and the possibility of scaling. These achievements not only improve morale but also enable businesses to earn the trust of new clients and partners.

In regard to support required for market expansion, entrepreneurs highlighted the need for systemic support. They demand additional awareness campaigns, education initiatives, and skill building in digital sectors to fill the knowledge gap among consumers. Export support, partnership, price subsidy, and promotion of local brands were also deemed to be of utmost significance. Education among the youth was most frequently proposed as the most crucial measure toward building a future consumer base that is sustainability-focused right from the early stages.

Lastly, the entrepreneurs suggested some recommendations to policymakers. They emphasize the need for heavy investment in digital infrastructure, especially in rural areas, and national eco-friendly campaigns and policies. Grants, SME loans, tax incentives, and marketing grants were viewed to be of highest priority. More special attention to women entrepreneurs, more trade shows, and partnership schemes were also suggested to create a successful sustainable business culture.

The insights show that while the sustainable business market is driven by strong purpose and growing consumer interest, it faces persistent challenges around cost, education, and trust. Authentic community engagement and storytelling have proven to be more effective than traditional marketing, and there is a clear call for holistic support—educational, financial, and infrastructural—to help these businesses thrive in the long run.

12.15 IMPLICATIONS

12.15.1 Managerial Implications

The research provides significant learnings for existing and potential entrepreneurs in the sustainable products industry. Perhaps, the most significant managerial lesson is to frame sustainable products not only as moral substitutes but also as value-based, high-quality products. Price remains a hindrance, so entrepreneurs need to develop strategies like cost-saving sourcing, creative pricing strategies (e.g., bundling, subscription), and explaining the long-term value of sustainable products. This involves highlighting durability, health advantages, and cost savings in the long run.

Another important implication is the need for establishing consumer trust and educating the marketplace. Managers need to emphasize clear, open, and consistent communication through storytelling, content marketing, product demonstrations, and third-party certification. Customer testimonials, influencer marketing, and user-generated content are also effective credibility builders. Sustainable entrepreneurs need to become not only peddlers but also teachers and evangelists.

Also, the research uncovers that successful companies tend to be deeply embedded in community and mission-branding. Managers should thus prioritize building community-oriented experiences like local workshops, eco-events, or ambassador programs that build emotional and cultural bonds with the brand. Strategic collaborations—whether with NGOs, schools, or other small companies—can boost reach and capital, making marketing more cooperative and affordable.

12.15.2 Social Implications

At a broader societal level, the results indicate that the sustainable business movement can contribute to behavior change, environmental awareness, and healthier communities. Nevertheless, there is still a huge gap in public awareness and accessibility. Sustainable consumption is regarded as a luxury, indicating the need for inclusive approaches to make green products more affordable and accessible to more people.

The research also highlights the contribution of social entrepreneurs to closing systemic gaps—be it addressing plastic pollution, food waste, or ethical sourcing. These companies are typically local solutions to international challenges, and their expansion adds directly to the sustainable development goals (SDGs). Thus, investing in such companies can have a compounding social effect: creating jobs, empowering communities, and protecting the environment.

A further social consequence is the necessity for early education and youth involvement on sustainability. Several entrepreneurs mentioned the necessity of reaching youth audiences to inculcate long-term environmentally conscious behaviors. Schools, community groups, and local governments must work together to develop programs that introduce young people and children to environmentally friendly ways of living, in the hope of inspiring upcoming consumers as well as agents of change.

12.15.3 Academic Implications

For researchers and academic institutions, the research provides various avenues for continued research and curriculum design. For academics, it underscores the increasing importance of sustainability entrepreneurship as a separate discipline worth further investigation. Researchers can look into more in-depth areas like consumer attitudes toward green products, the use of digital narratives in green branding, or the economic effects of grassroots green enterprises.

This research also offers a rich source of case studies and teaching materials for business schools. The challenges and strategies outlined here can be incorporated into marketing, entrepreneurship, and sustainability courses to enable students to learn about the actual complexities of operating a purpose-driven business.

In addition, the results indicate a significant relationship between individual motivation and entrepreneurial behavior, providing fertile terrain for psychological and behavioral research. Future studies can investigate how individual values are reflected in business models, or how narrative identity affects consumer participation in the sustainability arena.

Lastly, the research urges academia to bridge the gap between practice and research. Academic institutions can take a more proactive role in accommodating eco-entrepreneurs by offering mentorship, incubators, action-based research, and partnerships with local businesses. In so doing, not only do they enrich practical learning opportunities for students but also contribute to the resolution of actual issues of sustainability.

12.16 LIMITATIONS AND FUTURE IMPLICATIONS

In spite of recent research, significant gaps remain: First, limited attention to consumer attitudes toward women-owned sustainable brands. Second, lack of adequate research on comparative efficacy of digital versus conventional marketing techniques for sustainable products. Third, need for further investigation of policy frameworks that can enable improved market access and financing for women entrepreneurs. Future research should address these gaps to develop more targeted interventions and policy recommendations.

Interview Questionnaire: Marketing Challenges of Sustainable Products Among Women Entrepreneurs in Nigeria

Section A: Demographic Information

1. Age: _____
2. Educational Background: _____
3. Marital Status: _____
4. Location of Business: _____ (Urban/Rural)
5. Business Sector: _____ (e.g., fashion, agriculture, cosmetics etc.)
6. Business Ownership Type: _____ (Sole proprietorship/Partnership/Cooperative)

Section B: Business Background

7. Can you tell me about your business and the type of sustainable products you offer?
8. How long have you been in business?
9. What motivated you to start a business in sustainable products?
10. What is the size of your business (e.g., number of employees, revenue range, market reach)?

Section C: Marketing Challenges

11. What are the biggest challenges you face in marketing sustainable products in Nigeria?
12. How do consumers in your market perceive sustainable products? Do they understand the value of sustainability?
13. Do you face competition from non-sustainable products? If so, how do they impact your sales?
14. Have you encountered difficulties in setting competitive prices for your sustainable products?
15. What role do sociocultural factors play in marketing your products? Do traditional beliefs or consumer behaviors affect your sales?

Section D: Access to Resources and Support

16. What are your primary marketing channels (e.g., online, word-of-mouth, physical stores)? Which have been most effective?
17. Do you have access to financial resources to support marketing efforts? If not, how has this affected your business?
18. Have you received any support from the government, NGOs, or private institutions to promote sustainable entrepreneurship?
19. Do you use digital platforms (social media, e-commerce) to market your products? What challenges do you face in utilizing them?

Section E: Strategies and Opportunities

20. What strategies have you used to overcome marketing challenges? Have they been effective?

21. Are there any particular success stories or lessons learned from your experience in marketing sustainable products?
22. What opportunities do you see for improving the market for sustainable products in Nigeria?
23. What recommendations do you have for policymakers, investors, or support organizations to help women entrepreneurs succeed in this industry?

REFERENCES

Academia.edu. (2023). *Meeting the challenges of sustainable women entrepreneurship development in North-Eastern Nigeria.* Retrieved from https://www.academia.edu/87330555/Meeting_the_Challenges_of_Sustainable_Women_Entrepreneurship_Development_in_North_Eastern_Nigeria

Covenant University. (2023). *Women entrepreneurship and sustainability: A systematic literature review.* Retrieved from http://eprints.covenantuniversity.edu.ng/16978/1/Women%20Entrepreneurship%20and%20Sustainability.pdf

Etim, E. (2021). *The use of digital technologies for enhancing women entrepreneurship in Nigeria.* In Overcoming Challenges and Barriers for Women in Business and Education: Socioeconomic Issues and Strategies (pp. 45–66). IGI Global. https://doi.org/10.4018/978-1-7998-3814-2.ch003

EUDL. (2021). *Women entrepreneurship in Nigeria: Drivers, barriers and opportunities.* Retrieved from https://eudl.eu/pdf/10.4108/eai.7-6-2021.2308607

Munich Personal RePEc Archive. (2019). *Challenges of female entrepreneurship in Nigeria.* Retrieved from https://mpra.ub.uni-muenchen.de/93166/1/MPRA_paper_93166.pdf

Odunaike, B. A., & Ajiboye, O. E. (2020). *Low-income women entrepreneurs and household sustainability in Nigeria's border communities.* Journal of Women's Entrepreneurship and Education, 2020(3–4), 165–180. https://doi.org/10.28934/jwee20.34.pp165–180

Peter, W., & Orser, B. (2024). *Women entrepreneurs in rural Nigeria: Formal versus informal finance strategies.* International Journal of Gender and Entrepreneurship, 16(4), 602–621. https://doi.org/10.1108/IJGE-03-2023-0053

ResearchGate. (2023). *Women entrepreneurship and sustainable economic development: Evidence from South-Western Nigeria.* Retrieved from https://www.researchgate.net/publication/368328078_Women_Entrepreneurship_and_Sustainable_Economic_Development_Evidence_from_South-western_Nigeria

Sanni, M. A. (2024). *Role of inclusive formal financing infrastructure for women entrepreneurs in developing economies.* In Infrastructure Development Strategies for Empowering Women Entrepreneurs (pp. 300–318). IGI Global. https://doi.org/10.4018/979-8-3693-2917-7.ch014

Tandfonline. (2022). *Challenges hindering women entrepreneurship in Nigeria.* Retrieved from https://www.tandfonline.com/doi/full/10.1080/23311886.2022.2132675

Walden University. (2023). *Overcoming barriers to sustainability of female entrepreneurship in Nigeria.* Retrieved from https://scholarworks.waldenu.edu/cgi/viewcontent.cgi?article=11193&context=dissertations

Webster, J., & Watson, R. T. (2002). *Analyzing the past to prepare for the future: Writing a literature review.* MIS Quarterly, 26(2), xiii–xxiii.

Women's World Banking. (2024). *Empowering Nigeria's women micro-entrepreneurs: Actionable insights for financial services providers.* Retrieved from https://www.womensworldbanking.org/wp-content/uploads/2024/09/Empowering-Nigerias-Women-Micro-Entrepreneurs-Actionable-Insights-for-Financial-Services-Providers.pdf

13 Navigating Special Markets
Women Entrepreneurs in Special Industries

Mildred Amugune

13.1 INTRODUCTION

The landscape of entrepreneurship is changing rapidly, with women emerging as influential leaders in various fields. Historically, many specialized markets, those requiring specific knowledge, skills, or products have been predominantly male spaces. However, as society shifts and access to education broadens, more women are entering these niche areas and making their mark.

This chapter focuses on the experiences of women entrepreneurs within specialized industries. It highlights the various challenges these women face, the innovative strategies they use to overcome obstacles, and the unique strengths they contribute to their fields. Understanding these elements is essential, as it helps to identify ways to support and empower women in entrepreneurship.

While the growth of women-led businesses is evident across many sectors, their representation in specialized markets still requires attention. These markets often involve niche products and technical expertise, making them complex environments. Despite this, women are increasingly breaking through the barriers that have historically limited their participation.

By examining the distinct experiences of women entrepreneurs in these specialized industries, we can gain insights into the unique challenges they encounter. This knowledge is crucial for creating opportunities that foster growth and success for women in these fields. Through this exploration, we can better understand how to support and empower future generations of women entrepreneurs, paving the way for a more inclusive entrepreneurial landscape.

13.2 UNDERSTANDING SPECIALIZED MARKETS

Specialized markets are unique segments of the economy that focus on specific products and services tailored to meet particular needs. These markets stand apart from mainstream markets in several important ways.

DOI: 10.1201/9781003508953-13

13.2.1 Niche Products and Services

Specialized markets cater to niche demands, providing highly specific products or services. For example, a company may produce custom orthopedic shoes for individuals with specific foot conditions. Unlike general markets that offer a broad range of products, specialized markets focus on a particular audience and their unique preferences.

13.2.2 Technical Expertise

Engaging in specialized markets often requires advanced technical knowledge and skills. This necessity can pose challenges, especially for women who may encounter gender-based disparities in education and training related to STEM (Science, Technology, Engineering, and Mathematics) fields. Understanding these technical requirements is crucial for success in specialized industries.

13.2.3 Regulatory Complexity

Specialized markets are frequently governed by complex regulations, licensing requirements, and industry standards. These regulations can be overwhelming, particularly for women entrepreneurs who might lack access to the necessary networks and resources. Familiarity with regulatory frameworks is essential for navigating these markets effectively.

13.2.4 Customer Base

The customer base for specialized markets tends to be smaller yet more discerning. Customers in these markets often seek tailored solutions and expect high levels of service. This necessitates targeted marketing strategies and strong relationship-building efforts to meet the specific needs of customers, ensuring their satisfaction and loyalty.

13.2.5 Competitive Landscape

Competition within specialized markets can be fierce. Established companies often hold significant advantages over newcomers, including brand recognition, well-developed distribution networks, and established customer loyalty. To succeed, new entrants must develop innovative strategies to differentiate themselves and capture the interest of their target audience.

Understanding the characteristics of specialized markets is vital for anyone looking to thrive in these unique sectors. By recognizing the niche nature of these markets, the importance of technical expertise, the complexities of regulations, the discerning customer base, and the competitive landscape, entrepreneurs and businesses can better navigate the challenges and opportunities within specialized markets.

13.3 CHALLENGES FACED BY WOMEN ENTREPRENEURS IN SPECIALIZED MARKETS

Women entrepreneurs play a vital role in the economy, bringing innovation and diversity to various industries. However, their journey in specialized markets often involves significant challenges that can hinder their success and growth. Understanding these obstacles is crucial for fostering an environment that supports women in entrepreneurship.

13.3.1 Access to Capital

One of the most pressing challenges faced by women entrepreneurs is gaining access to financial resources. Research shows that women often encounter disparities in securing funding compared to their male counterparts. Traditional lending institutions may have inherent biases, leading to lower approval rates for women-led ventures. Furthermore, many women entrepreneurs lack the robust financial networks that facilitate investment opportunities, making it difficult to obtain the necessary capital to launch or expand their businesses.

13.3.2 Gender Bias and Stereotypes

Societal perceptions regarding gender roles can significantly impact women's entry into specialized markets. Stereotypes often associate technical expertise and leadership skills with men, which can undermine the confidence of women entrepreneurs. This bias is particularly evident in fields such as technology and engineering, where women may struggle to be recognized as capable leaders and innovators.

13.3.3 Balancing Responsibilities

Women frequently bear a disproportionate share of domestic responsibilities, which can limit their ability to dedicate time and energy to their entrepreneurial endeavors. The challenge of balancing family obligations with the demands of running a business can create substantial stress and impede growth opportunities. As a result, many women entrepreneurs find themselves stretched thin, trying to juggle multiple roles.

13.3.4 Networking Barriers

Networking is essential for success in specialized markets, yet women often face significant barriers when trying to establish connections in male-dominated industries. The lack of access to influential networks can restrict opportunities for mentorship, partnerships, and collaboration, further isolating women entrepreneurs. This isolation can hinder their ability to navigate the complexities of their industries and access valuable resources.

13.3.5 Market Entry Barriers

Women entrepreneurs in specialized markets also encounter various market entry barriers. Navigating complex regulatory frameworks, building strong industry relationships, and establishing a credible brand can be particularly challenging. These hurdles can stifle innovation and limit the potential for growth within their businesses.

While women entrepreneurs contribute greatly to the economy, their path in specialized markets is often fraught with challenges. Addressing these obstacles such as access to capital, gender bias, balancing responsibilities, networking barriers, and market entry challenges is essential to create an equitable entrepreneurial landscape. By understanding and actively working to overcome these challenges, we can empower women entrepreneurs to thrive and succeed in their respective fields.

13.4 STRATEGIES FOR WOMEN TO THRIVE IN SPECIALIZED MARKETS

Women entrepreneurs face unique challenges when navigating specialized markets, but with the right strategies, they can achieve significant success. This chapter outlines effective approaches that empower women to thrive in these competitive environments.

13.4.1 Building Strong Networks

Creating and nurturing a robust network is essential for women entrepreneurs. Engaging with organizations focused on women's entrepreneurship provides valuable resources, mentorship, and support. Networking with other female entrepreneurs allows women to share experiences and solutions, fostering a community that understands the unique challenges they face. Forming connections with industry experts can also open doors to new opportunities and insights.

13.4.2 Seeking Education and Training

Investing in education and skill development is crucial for women looking to enter specialized markets. Access to training programs, workshops, and courses equips women with the technical knowledge and confidence needed to succeed. Continuous learning helps women stay updated on industry trends, technological advancements, and regulatory changes, ensuring they are well-prepared to navigate their sectors effectively.

13.4.3 Leveraging Technology

The rise of digital platforms has created new opportunities for women entrepreneurs. Utilizing e-commerce, social media, and online marketing allows women to reach broader audiences and establish their brands without the constraints of traditional market entry barriers. Technology can streamline operations, improve customer engagement, and reduce costs, enabling women to operate more efficiently and effectively.

13.4.4 Developing Strong Partnerships

Collaborating with other businesses, industry associations, and government agencies can help women access vital resources. Forming strategic partnerships enhances credibility and expands market reach. By working together, women entrepreneurs can share resources, knowledge, and support, ultimately driving mutual success in specialized markets.

13.4.5 Embracing Continuous Learning

Staying up-to-date with industry trends and innovations is vital for success. Women entrepreneurs should embrace a mindset of lifelong learning, seeking out opportunities for professional development and personal growth. By doing so, they can adapt to changes in their markets and remain competitive.

13.4.6 Advocating for Policy Change

Engaging in advocacy efforts to address systemic barriers is crucial for women entrepreneurs. Collaborating with organizations that promote gender equity in business can lead to policy changes that support access to capital, networking opportunities, and resources tailored to women-led enterprises. Through collective action, women can create a more equitable environment for future generations of entrepreneurs.

13.4.7 Highlighting Unique Strengths

Women entrepreneurs can leverage their unique perspectives, empathy, and collaborative skills to differentiate themselves in the market. By embracing their strengths, they can create innovative solutions and foster inclusive business practices that resonate with diverse audiences.

By employing these strategies, women can navigate challenges effectively and achieve lasting success in their entrepreneurial endeavors.

13.5 UNIQUE STRENGTHS OF WOMEN ENTREPRENEURS IN SPECIALIZED MARKETS

Women entrepreneurs play a vital role in specialized markets, bringing a unique set of strengths that significantly enhance their impact and contribution. Their skills and perspectives not only foster innovation but also drive social change, making them formidable players in the business world.

13.5.1 Innovation and Creativity

One of the standout qualities of women entrepreneurs is their ability to innovate and think creatively. They approach problems with fresh perspectives, leading to the development of unique products and services that can fill gaps in the market. This creativity often results in solutions that are not only effective but also resonate with consumers on a personal level.

13.5.2 Strong Interpersonal Skills

Women are known for their exceptional communication and interpersonal skills. These strengths allow them to build strong relationships with customers, suppliers, and partners, creating networks that are essential for business success. Their ability to connect with others fosters trust and loyalty, which can be particularly advantageous in specialized markets where customer relationships are key.

13.5.3 Social Impact

Many women entrepreneurs are motivated by a desire to create a positive social impact. This drive often leads them to establish businesses that prioritize sustainability and ethical practices. By focusing on social responsibility, they not only contribute to their communities but also appeal to consumers who are increasingly concerned about the ethical implications of their purchases.

13.5.4 Resilience and Adaptability

Resilience is another defining characteristic of women entrepreneurs. They frequently face challenges and barriers in their entrepreneurial journeys but demonstrate remarkable perseverance. Their ability to adapt to changing market conditions and overcome obstacles is a significant asset, particularly in specialized markets where unpredictability is common.

13.5.5 Diverse Perspectives

Women bring diverse experiences to their industries, which enhances creativity and innovation. This diversity allows for the development of products and services that meet the needs of a broader consumer base. By incorporating various viewpoints, women entrepreneurs can create offerings that resonate with a wider audience.

13.5.6 Empathy and Relationship-Building

Women often excel in empathetic leadership and relationship-building. These qualities enhance collaboration and improve team dynamics, which are crucial in specialized markets. Understanding customer needs and fostering strong partnerships are vital for success, and women entrepreneurs often excel in these areas.

13.5.7 Commitment to Sustainability

A commitment to sustainable practices is common among women entrepreneurs. They often focus on creating ethical and environmentally friendly products, which not only benefits the planet but also aligns with consumer values. As sustainability becomes a priority for many consumers, women entrepreneurs who prioritize these practices can position themselves favorably in specialized markets.

The unique strengths of women entrepreneurs, including their creativity, interpersonal skills, social impact, resilience, diverse perspectives, empathetic leadership, and commitment to sustainability, make them invaluable contributors to specialized markets. Their ability to innovate and adapt not only enhances their businesses but also positively influences the communities they serve.

13.6 CONCLUSION

The landscape of women's entrepreneurship in specialized markets is evolving, revealing both challenges and opportunities. As women continue to navigate these sectors, their unique strengths, innovative approaches, and commitment to overcoming barriers will play a pivotal role in shaping the future of entrepreneurship. By recognizing the distinct characteristics of specialized markets and implementing strategies to thrive, women entrepreneurs can not only succeed personally but also contribute to a more diverse and inclusive entrepreneurial ecosystem.

This chapter underscores the importance of supporting women in their entrepreneurial journeys, ensuring that the path to success in specialized markets is accessible and equitable for all.

Index

S

T

U

V

W

For Product Safety Concerns and Information please contact our EU representative GPSR@taylorandfrancis.com Taylor & Francis Verlag GmbH, Kaufingerstraße 24, 80331 München, Germany